THE TRAVEL BOOK

A JOURNEY THROUGH
EVERY COUNTRY IN THE WORLD

The Story of The Travel Book

Most travel journeys take in just a country or two, but the journey on which you're about to embark incorporates every country on Earth. In this book you'll find evocative glimpses of every nation, from Afghanistan to Zimbabwe, from the postage-stamp-sized Vatican City to the epic expanse of Australia.

To visit all the countries in this book would require several passports, or it can be simulated with a turn of these pages. Highlighted by fascinating photography, *The Travel Book* offers a glimpse of each country's perks and quirks, suggesting ways to immerse yourself in their life and their landscapes. What results is a snapshot of our kaleidoscopic world rather than an encyclopedic reference. It's a book that unashamedly views the planet through the prism of the traveller.

In this new edition, we have celebrated each country's achievements and diversity. We hope this ethos of positivity inspires your own get-up-and-go spirit. Travel faces its own challenges, from security to sustainability. But, if approached responsibly, experiencing other cultures will benefit not only ourselves but host and home countries.

The Country Conundrum

How do you define a country? By most criteria, a country is an area of land with defined borders and a population that identifies itself as being from that country. It must have a government or ruler and also be recognised by other countries and international bodies as an independent nation. As a starting point, we've used the United Nations' list of 193 member states. Every one of these countries features in the book.

And then there are less clear-cut entries. We have included the four constituent countries of the United Kingdom – England, Northern Ireland, Wales and Scotland – each of which has a form of self-government. Kosovo and Taiwan also meet many of these criteria, although they are not recognised as countries by the UN. Antarctica, portions of which are claimed by several different countries, including Australia, New Zealand and Norway, was too large and extraordinary to leave out.

But we have not included overseas territories, such as many of the Caribbean islands that were once part of former European empires. French Guiana, for example, is an overseas department of France and therefore a member of the European Union in South America. There are many different types of territory: Greenland is an autonomous territory of Denmark while Tokelau is a dependent territory of New Zealand. Puerto Rico has been an unincorporated territory of the USA since 1898. And the Cook Islands are described as a 'territory in free association with' New Zealand. The short answer as to how you define a country or a territory is 'it's complicated'.

The Structure

The Travel Book follows the most straightforward of formats – A to Z – rolling through the alphabet of nations. From a traveller's perspective, a country's might isn't necessarily relative to its fascination and appeal, and we've tried to capture that, giving equal weight to every country regardless of its population size or land mass.

The book's guiding philosophy is to present a subjective view of the world from Lonely Planet's perspective, looking below the surface to show a slice of life from every country in the world. Entries evoke the spirit of each place by appealing to the senses – what you might see and feel, what kind of food and drink might flavour your visit, and which books, music or films will help prepare you for the experience. You'll find the events, objects and people that are central to each country's identity and you'll find curious, little-known facts.

You may never visit all *The Travel Book*'s destinations, but if it's true, as Aldous Huxley once wrote, that 'to travel is to discover that everyone is wrong about other countries', then to read about them all is to find out if you are right.

1. The tribal government of the Blackfeet Nation is on this reservation in Browning, Montana, USA

2. Taking a bracing midnight swim in the Lemmenjoki River, which flows through Finnish Lapland

3. Wander between tiny bars and *izakayas* in Tokyo's alleys for a perfect night out

4. Arles' Roman amphitheatre in the south of France once held 20,000 spectators for chariot races

5. A superb starling on the savannah of the Serengeti in Tanzania

6. Crossing New Zealand's Waitakere Ranges from Auckland to Piha in a campervan

Best time to visit

April to June and September to October – with all
visits dependent on the political weather

Top things to see

- Signs of normality emerging in once war-torn
  Kabul, including the revitalised National Museum
- The dizzying 800-year-old Minaret of Jam
- The skyline of medieval Herat, punctuated by
  bristling minarets
- The blue domes of the Shrine of Hazrat Ali in
  Mazar-e Sharif, Afghanistan's holiest pilgrimage
  site
- The Panjshir Valley, with its rushing river and
  immaculate villages and orchards

Top things to do

- Contemplate the holes formerly occupied by
  giant Buddha statues in the Bamiyan Valley
- Trek with yaks across the Pamir Mountains in the
  Wakhan Corridor
- Dip your toes in the blue mineral waters of the
  Band-e Amir lakes
- Feel a sense of peace in Babur's rehabilitated
  Mughal gardens in Kabul
- Haggle for Afghan carpets at their source with
  nomadic traders

Getting under the skin

Read: Eric Newby's witty *A Short Walk in the Hindu
Kush*; Rory Stewart's *The Places In Between*, an
excellent post-Taliban travelogue
Listen: to everything from spiritual *qawalis* (Sufi
religious chants) to the political rap of Sayed Jamal
Mubarez – something unthinkable during the
music-free years of the Taliban
Watch: *Osama*, directed by Siddiq Barmak,
following a girl who assumes a male identity to
work in Taliban-era Kabul
Eat: fat Kandahari pomegranates and (according to
Marco Polo) the best melons in the world
Drink: *chai sabz* (green tea) scaldingly hot at a
traditional teahouse

In a word

Salaam aleikum (Peace be with you)

Celebrate this

Women's rights are inching forwards in
Afghanistan, thanks to the efforts of women
such as 70-year-old Sakena Yacoobi, who has
campaigned for education for girls under shahs,
warlords and the Taliban.

Random fact

The lapis lazuli gemstones used to make
Tutankhamun's death mask were mined in
northeastern Afghanistan.

1. The Great Mosque
   of Herat is largely
   unchanged since the
   15th century

2. Every culture
   leaves something in
   Afghanistan; surfers
   catching waves in the
   Panjshir Valley

3. A player on the Afghan
   women's football team
   trains in a Kabul gym

A CAPITAL KABUL // POPULATION POPULATION 36.6 MILLION // AREA 652,230 SQ KM // OFFICIAL LANGUAGES DARI & PASHTO

Afghanistan

Today, only the most hardened travellers brave the fragile security situation to reach Afghanistan's hidden treasures, but for much of its history, this was a place of peace and tranquillity. In the first century BCE, the foothills of the Hindu Kush were home to a Buddhist civilisation of rare artistic refinement. Later, Islamic Kabul was an essential stop on the hippy trail to India. The politics of the 19th-century 'Great Game' between Britain and Russia sowed the seeds for centuries of geopolitical turmoil, with Afghanistan caught in the crossfire. Yet despite this backdrop of conflict, the old Afghanistan endures: mosques and fortresses straight out of *The Road to Oxiana*, ancient ruins that speak of centuries at the crossroads of Asia, and a culture as timeless as the desert itself.

A CAPITAL TIRANA // POPULATION 3 MILLION // AREA 28,748 SQ KM // OFFICIAL LANGUAGE ALBANIAN

Albania

After decades of isolation from the rest of Europe under harsh communist rule, Albania emerged in the early 1990s, blinking, as it finally entered the modern age, and it has been on a rollercoaster ride of change ever since. Endowed with incredible natural beauty, including some of Europe's most impressive wilderness in the Accursed Mountains and miles of beaches, the 'land of the eagles' is no longer a secret among backpackers wanting to escape the crowds elsewhere in Europe. From the (literally) colourful capital Tirana to its old Ottoman towns, Albania beckons with remote hiking trails, fresh seafood and strong raki, thought-provoking museums and much more.

Best time to visit
April to October

Top things to see
- Historic Berat, a near-perfectly preserved Ottoman town of whitewashed houses climbing up a hillside
- The serene ruins of ancient Butrint, lost deep in the forest with a lakeside setting
- Sprawling and fast-paced capital Tirana, full of museums, cafes and shopping
- The mountaintop town of Gjirokastra, with a looming castle and grand Ottoman mansions

Top things to do
- Discover your own beachside idyll on the many beaches and coves of the Ionian coast
- Jump aboard the daily ferry on Lake Koman for a trip into Albania's mountainous interior
- Do the exhilarating day trek between the villages of Valbona and Theth
- Plunge into the bottomless, glassy depths of the Blue Eye Spring

Getting under the skin
Read: *Chronicle in Stone* by Ismail Kadare, a boyhood tale set in Gjirokastra; *Crossing* by Kosovo-born Pajtim Statovci, set in post-communist Albania
Listen: to the entwined vocal and instrumental parts of traditional southern Albanian polyphony
Watch: Bujar Alimani's *Amnesty*, a drama about two spouses making conjugal visits to a prison together
Eat: Roast lamb in the mountains or freshly caught fish along the coast; or *byrek*, the quintessential Albanian fast food (layered pastry filled with cheese, potato or minced meat)
Drink: raki (aniseed-flavoured grape brandy) as an aperitif; or *konjak* (cognac) as an after-dinner tipple

In a word
Tungjatjeta (Hello)

Celebrate this
Albania's bunkers, the Cold War legacy of communist dictator Enver Hoxha, are found across the country. Recently, an enormous bunker in the capital has been transformed into an excellent history and contemporary art museum: Bunk'Art.

Random fact
The Albanian language is unrelated to any other in Europe and is thought to derive from ancient Illyrian.

1. Valbona Valley National Park in northern Albania

2. Ksamil beach in Butrint National Park in the south of Albania

3. The 18th-century bridge of Mesi, near Shkodra

3

Best time to visit
November to April

Top things to see
- The Islamic heart of Algiers, secreted in the city's steep hillside Casbah
- Djemila's Roman ruins, beautifully sited in the Mediterranean's hinterland
- The stunning pastel-coloured oasis towns in the M'Zab Valley on the Sahara's northern fringe
- A spectacular sunrise from atop barren mountains deep in the Sahara at Assekrem
- Tassili N'Ajjer's open-air gallery of rock art from the time before the Sahara became a desert

Top things to do
- Sip a cafe au lait in Algiers' French-style sidewalk cafes then dive into the Casbah
- Dream of Algeria's Roman and Phoenician past at the charming old port of Tipaza
- Discover the hidden treasures of Tlemcen's extraordinary Arab-Islamic architecture
- Sleep among the sand dunes of the Grand Erg Occidental
- Explore the Tassili du Hoggar, with some of the Sahara's most beautiful scenery

Getting under the skin
Read: *What the Day Owes the Night* by Yasmina Khadra (translated by Frank Wynne), an incredible story following a young boy in 20th-century French Algeria; Jeremy Keenan's *Sahara Man: Travelling with the Tuareg*, which takes you deep into the Tuareg world

Listen: to *King of Rai: The Best of Khaled* for Algeria's best-loved musical (and most danceable) export

Watch: Mounia Meddour's *Papachia*, a semi-autobiographical tale of a design student confronting life during the country's civil war

Eat: chickpea fritters, couscous and lamb tajine spiced with cinnamon

Drink: three servings of strong tea around a Tuareg campfire

In a word
Salaam aleikum (Peace be with you)

Celebrate this
Nabila Djahnine: she fought tirelessly for the rights of Algerian women until her assassination in 1995.

Random fact
At 2.4 million sq km (almost 900,000 sq km larger than Western Europe), Algeria is Africa's largest country.

1. Algiers' Casbah, with its citadel, is a Unesco World Heritage Site

2. The Arch of Caracalla at Djemila; the city was abandoned after the fall of the Roman Empire in the 5th and 6th centuries

3. The ancient sandstone formations of Assekrem in the Hoggar mountains

A

CAPITAL ALGIERS // **POPULATION** 43.0 MILLION // **AREA** 2,381,740 SQ KM // **OFFICIAL LANGUAGES** ARABIC & TAMAZIGHT

Algeria

Stretching from the fertile, mountain-backed shores of the Mediterranean to the heart of the Sahara, Algeria's landscape has been home to many empires and dynasties over several millennia. Battles over control have continued to be waged, most recently with the civil war of the 1990s. While subsequent peace has opened the country to travel, insecurity in border regions and remote areas of the Sahara still poses a challenge. Those who explore will find a catalogue of attractions – from some of the world's best-preserved Roman cities in the north to the extraordinary Saharan landscapes and oasis towns of the south.

A CAPITAL ANDORRA LA VELLA // POPULATION 77,000 // AREA 468 SQ KM // OFFICIAL LANGUAGE CATALAN

Andorra

Racing down snowy pistes, sleeping snug between ice-hotel walls and shopping in some of Andorra's hundreds of low-tax stores is how most think of this principality, neatly wedged between France and Spain in the eastern Pyrenees. Fewer know about its history (which harks back to the 9th century), its fascinating people, the cobbled core of Andorra la Vella (the country's only town) or its secret hoard of thermal spas that soothe weary limbs. The absence of an airport means Andorra never gets the attention it deserves, but jet into Barcelona in neighbouring Spain or Toulouse in France and this intriguing mini-nation is just a couple of hours' drive away.

Best time to visit
Mid-December to early April for skiing; June to September for hiking

Top things to see
- The cobbled streets and hidden squares of Andorra la Vella's quaint historic quarter
- Three near-pristine valleys – each justifies a one-day hike, at least
- The stunning mountain scenery of Grandvalira, the largest ski area in the Pyrenees
- The Museu del Tabac in Sant Julià de Lòria, a fascinating tobacco-factory-turned-museum
- Historic Casa de la Vall in Andorra la Vella, built in 1580 as a home for a wealthy family, now housing Andorra's parliament

Top things to do
- Ski the winter slopes of Soldeu in the heart of Grandvalira
- Wallow in toasty-warm mineral water at Europe's largest spa complex, Caldea, in Andorra la Vella
- Join the crowds at dusk on the rooftop of Plaça del Poble, Andorra la Vella, oohing and aahing over valley views
- Hike the 112km Grand Route de Pays (GRP): seven days of hut-to-hut tramping across the roof of Europe

Getting under the skin
Read: *A Tramp in Spain: From Andalusia to Andorra*, a travelogue by Englishman Bart Kennedy, who tramped, knapsack on back, to Andorra in 1904
Listen: to something classical by the National Chamber Orchestra of Andorra directed by top Andorran violinist Gerard Claret
Watch: *Nick* (Outlier) a thriller from Spanish director Jose Pozo about a teen trying to solve a murder he witnesses in the mountains of Andorra
Eat: hearty mountain fare – *trinxat* (bacon, potatoes and cabbage) or traditional *escudella* (a warming chicken, sausage and meatball stew)
Drink: mulled red wine laced with lemon, apple, raisins, cinnamon and cognac after a day on the slopes

In a word
Hola (Hello)

Celebrate this
With barely any sizeable urban centres Andorra is all about enjoying nature, making hiking, cycling, climbing and a host of other outdoor activities the highlight of a visit.

Random fact
In the 1980s a 2km glacier still slumbered in the Andorran Pyrenees; today it's completely gone and glacial lakes are all that remain.

1. Arinsal village is a hub for hikers in summer and skiers in winter

2. Caldea thermal spa uses water from one of Andorra's 35 natural springs

3. Pyrenean hiking trails surround the Estanys de Tristaina mountain lakes

1. Catching waves at Cabo Ledo in Bengo Province; tuition and guiding is available

2. The pass at Serra da Leba near Lubango

3. Notice the colonial Portuguese influence in Namibe Province

A CAPITAL LUANDA // POPULATION 32.5 MILLION // AREA 1,246,700 SQ KM // OFFICIAL LANGUAGE PORTUGUESE

Angola

Angolans are fighters – but they are lovers, too. The latter fact has long been lost on the world's press, who have labelled this African nation a haven of havoc, broadcasting stories of its civil unrest, blood diamonds, wasted oil revenues, sky-high cost of living and starving people. However, for those who successfully navigate the red tape required to land a rare tourist visa, it's the love that they'll remember. Whether it's an unquestioning love of God, an enthusiastic embrace of romance or an unwavering desire to dance like there is no tomorrow, the passion of Angolans is intoxicating to observe.

Best time to visit
June to September during the cooler dry season

Top things to see
- The cascading torrents of Kalandula Falls, one of Africa's largest waterfalls
- The crumbling art deco and neoclassical facades in the coastal town of Namibe
- Miradouro de Lua, a rusty-topped Martian-like rock formation that drops to the Atlantic
- The markets, volcanic fissures and giant statue of Christ in the city of Lubango

Top things to do
- Climb the dunes in Parque Nacional do Iona
- Spot the replenished large fauna within Parque National da Kissama
- Grab a surfboard and ride the cool Atlantic swells; they are some of Africa's best
- Take the Benguela railway from Lobito to the Congo's fringes

Getting under the skin
Read: *Blue Dahlia, Black Gold,* which follows author Daniel Metcalfe on his travels as he quests to put his finger on the pulse of this palpitating nation
Listen: to the romantic rhythms of *kizomba*, an evolution of the country's traditional *semba* music
Watch: *Another Day of Life*, an award-winning animated feature based on Polish reporter Ryszard Kapuscinski's journey through Angola in 1975
Eat: *calulu de peixe* (fish stew)
Drink: *galãos* (white coffee) – Angola has historically been one of the world's largest producers of the bean, and locals love drinking it with milk

In a word
Tudo bom? (How's things?)

Celebrate this
Musician Toty Sa'Med: he works to keep the Angolan language of Kimbundu alive with his lyrics.

Random fact
Although polygamy is illegal, it was common after the civil war due to a shortage of men.

Best time to visit
November to February for 'summer'; March for marine mammal spotting

Top things to see
- Colossal icebergs and mountain reflections on looking-glass waters at Paradise Harbor
- The true grit of Antarctic exploration icily preserved at Shackleton's expedition hut
- Whales moving in for the krill at heart-breakingly beautiful Wilhelmina Bay
- Thunderous glaciers calving and penguins skittering across the ice at Neko Harbor on the continent proper
- Penguins gracefully porpoising through the sheer-sided Lemaire Channel

Top things to do
- Glide on a Zodiac under the morning's pink skies past basking Weddell seals and honking gentoo penguins
- Sail into the caldera of a still-active volcano and hang out with chinstrap penguins on Deception Island
- Gasp at the echoing 'pffft' of a humpback surfacing next to your boat then marvel at acrobatic displays of lobtailing and breaching
- Experience sensory overload – the tang of salt, the biting cold and the pitter-patter of a thousand happy penguin feet

Getting under the skin
Read: *Endurance*, a gripping account of Sir Ernest Shackleton's epic, ill-fated voyage to cross the Antarctic continent in 1916
Listen: to *The Antarctic Sun* podcast going behind the scenes of the United States Antarctic Program
Watch: Sir David Attenborough's six-part BBC documentary series, *Frozen Planet* (2011)
Eat: an Antarctic barbecue, set up on deck or even on the ice
Drink: an Antarctic Old Fashioned: a fruity mix of 100-proof bourbon, multiflavoured Life Savers sweets and just-melted snow

In a word
The A-factor (The local term for the unexpected difficulties caused by the Antarctic environment)

Celebrate this
Following the inauguration of US President Joe Biden in 2021, the return of the United States to the Paris Agreement of the United Nations Framework Convention on Climate Change renews hope to meet the global challenges faced by the planet's polar regions.

Random fact
Antarctica's ice sheets contain 90% of the world's ice – 28 million cu km – or about 70% of the world's fresh water.

1. Emperor penguins nurture their chicks on Snow Hill Island

2. Ernest Shackleton sheltered in this hut on Cape Royds during his 1907-09 expedition

3. Giant icebergs glide along Lemaire Channel

THE TRAVEL BOOK

A POPULATION 4490 (SUMMER), 1106 (WINTER) // AREA 14.2 MILLION SQ KM

Antarctica

Snow, icebergs the size of multistorey car parks, wave-lashed seas, knife-edge mountains, ever-changing light. Stark and staggeringly remote, Antarctica and the enormousness of its ice shelves and mountain ranges make for a haunting, elemental beauty. The wildlife rules here, and breathtakingly close encounters with gentoo, Adélie and chinstrap penguins, seals and humpback whales are common. Governed by 29 nations, this continent is primarily dedicated to scientific research but it's also one of the planet's regions that's most visibly affected by climate change; as parts of the continent melt, others grow, leaving scientists to debate what it all actually means.

1. Darkwood Beach on Half Hyde Bay is one of Antigua's finest

2. English Harbour, with Falmouth in the distance, was Horatio Nelson's dockyard

3. Fig Tree Drive features mainly coconut palms

A CAPITAL ST JOHN'S // POPULATION 96,286 // AREA 441 SQ KM // OFFICIAL LANGUAGE ENGLISH

Antigua & Barbuda

Antigua is a bustling island that's home to many a commercial business and a fair number of wintering celebrities, such as Eric Clapton, Ken Follet, Oprah and more. Locals claim that the island has 365 beaches, one for every day of the year. While that might be a tough claim to verify, these strands of sand are a sure bet. Meanwhile Antigua's tiny sibling Barbuda has a fraction of the population, all scattered about a paradise of pink and white sand, much of it accessible only by boat.

Best time to visit
December to mid-April

Top things to see
- Half Moon Bay, where crystal blue waters lap a white crescent in the southeast of Antigua
- Frigate Bird Sanctuary, one of the world's largest colonies of frigate birds
- Fig Tree Drive winding through rainforest, past roadside stands and dotted with hiking trails
- St John's Public Market, filled with local foods and the perfect place to grab a beach picnic

Top things to do
- Hop on a catamaran to cruise the coastline and snorkel Cades Reef
- Hit the beach on the southwest coast at sublime Ffryes, Turner's, Darkwood or Jolly Bay
- Enjoy Sunday afternoon barbecue at Shirley Heights Lookout Restaurant, swaying your hips to steel drum and reggae

Getting under the skin
Read: Jamaica Kincaid's novel *Annie John*, a story of growing up in Antigua
Listen: to steel pan; calypso; ubiquitous reggae music; and zouk, the party music
Watch: *The Sweetest Mango* – Antigua's first feature film is a sweet romantic comedy
Eat: *duckanoo* (a dessert made with cornmeal, coconut, spices and brown sugar); or black pineapple, sold along Fig Tree Drive
Drink: locally brewed Wadadli beer; or Antiguan rums Cavalier or English Harbour

In a word
Cool out (take life easy)

Celebrate this
The islands' food scene is flourishing with interesting takes on traditional dishes. Check out roadside stands: look for those with the longest queues.

Random fact
Most of Barbuda's 1600 people share half a dozen surnames, descended from enslaved people.

Best time to visit
March to May (autumn) for Buenos Aires, December to March (summer) in Patagonia

Top things to see
- The soaked and ear-shattering panorama of the spectacular Iguazú Falls
- The sultry steps of tango performed on the streets
- Endangered southern right whales in Reserva Faunística Península Valdés
- The massive Glaciar Perito Moreno calve with thunderous cracks
- Stray cats and sculpted mausoleums in Recoleta Cemetery, the final stop for Buenos Aires' rich and famous

Top things to do
- Live it up till the sun comes up in Buenos Aires' bars and clubs
- Feel sporting tensions run high at a Superclásico clash between Buenos Aires footballing heavyweights Boca Juniors and River Plate
- Test-run various vintages on a *bodega* (winery) tour outside Mendoza
- Ride the windy range with *gauchos* (cowboys) at a Patagonian *estancia* (ranch)
- Feast on steaks and every other part of a cow at a backyard *asado* (barbecue)

Getting under the skin
Read: anything by the lyrical master of the short story, Jorge Luis Borges
Listen: to the trap-influenced hip-hop of Paulo Londra, the tangos of Carlos Gardel and the gritty classic Argentine rock of Charly García
Watch: a young Che Guevara discover Latin America in *The Motorcycle Diaries* or quirky man-and-his-dog road trip flick *Bombón: El Perro*
Eat: *empanadas* (pastries stuffed with savoury fillings), *alfajores* (delicious multilayered cookies) and *facturas* (sweet pastries)
Drink: maté (pronounced mah-tay), a bitter tea served in a gourd with a metal straw and shared among friends and colleagues; or a fruity Malbec for something stronger

In a word
Qué copado! (How cool!)

Celebrate this
In 2009, in remote Tierra del Fuego, Alex Freyre and Jose Maria Di Bello became South America's first same-sex married couple. The following year, Argentina became the first country in South America (and the second in the Americas, after Canada) to legalise same-sex weddings.

Random fact
Football may be the country's obsession, but the national sport is actually *pato*, which mixes elements of basketball and polo and was once played with live ducks.

1. Cementerio de la Recoleta has more than 6000 graves including that of Eva Perón

2. Gauchos herding cattle

3. The road to El Chalten in Los Glaciares National Park

CAPITAL BUENOS AIRES // **POPULATION** 45.5 MILLION // **AREA** 2,780,400 SQ KM // **OFFICIAL LANGUAGE** SPANISH

Argentina

Think big. Argentina boasts the highest Andean peak (Aconcagua) and the world's southernmost city (Ushuaia), while the epic stretch of Patagonia passes glaciers and mountains on its way to the southern seas. Buenos Aires offers countless opportunities for all-night revelry, while beyond the city limits, nature comes unabashed and boundless. The dry pastel hues of the northern desert erupt into the thunderous falls of Iguazú, the crisp skies of the lakes region and the craggy wilderness of the south. Do as locals do: slow down, accept time is fluid, and you will draw in all manner of encounters.

A CAPITAL YEREVAN // POPULATION 3 MILLION // AREA 29,743 SQ KM // OFFICIAL LANGUAGE ARMENIAN

Armenia

At the meeting point of Europe and the Middle East, the high rocky plateau of Armenia is a redoubt of artistry and remote beauty. Armenians experienced an often traumatic 20th century. Yet they have endured, seeking solace in their Christian faith, and maintaining a fierce pride in their language, culture and homeland. Flaunting mountain views, monasteries, walking trails, sublime stone architecture and the lively city of Yerevan, Armenia offers an experience steeped in culture, outdoor adventure and the timeless pleasure of getting to know an ancient, cultured and little-known nation.

Best time to visit
May to September

Top things to see
- Yerevan, the cultural heart of the nation, with fabulous museums, galleries and the buzzing Vernissage flea market
- Holy Echmiadzin, the seat of the Armenian Apostolic Church
- The artists' retreat of Dilijan, full of gingerbread-style houses, fresh mountain air and hiking paths along a section of the Transcaucasian Trail
- Vayots Dzor, the southern province peppered with monasteries, walking trails, the Selim caravanserai and wine-growing Arpa Valley

Top things to do
- Slow down and tap into the fabulous street culture of Yerevan
- Descend to the snake pit once occupied by St Gregory the Illuminator, in Khor Virap Monastery
- Trundle through the Debed Canyon, with forested valleys, quiet villages and monasteries
- Spend the day in Goris, home of potent fruit brandies and 5th-century cave houses

Getting under the skin
Read: *Visions of Ararat* by Christopher Walker, a compilation of writings by Armenian soldiers, anthropologists and poets; or *Family of Shadows* by Garin Hovannisian, a masterful blend of family and national history
Listen: to *Black Rock*, full of the mournful melodies of the *duduk* (traditional double-reed flute) played by master Djivan Gasparyan
Watch: the poetic *Colour of Pomegranates* by Sergei Paradjanov; or the soapy historical romance *The Promise*, set during the Armenian genocide and starring Oscar Isaac and Christian Bale
Eat: *khoravats* (skewered pork or lamb): lighting the barbecue is almost a daily ritual
Drink: *soorch* (gritty and lusciously thick coffee); or *konyak* (cognac), the national liquor

In a word
Genats (Cheers!)

Celebrate this
Few nations excel at chess more than Armenians, so much so that chess is a compulsory school subject for six to eight year olds.

Random fact
Armenia was the first nation to accept Christianity as a state religion, converting en masse in 301 CE.

1. The twin peaks of Mt Ararat overlook Yerevan

2. Building began at Haghpat Monastery in the 10th century

3. The Armenian Genocide Memorial stands on Tsitsernakaberd in Yerevan

Australia

Australia is as big as your imagination. There's a heckuva lot of tarmac across this wide brown land and the best way to appreciate the country is to hit the road. Sure, it's got deadly creepy crawlies and sharks, but they don't stop people from coming here to see its famous natural beauty – from endless sunbaked deserts to lush tropical rainforest and wild southern beaches. Scattered along the coasts, its cosmopolitan cities blend an enthusiasm for art and food with a passionate love of sport and the outdoors. Those expecting to see an opera in Sydney one night and spy crocs in the Outback the next morning will have to rethink their geography: it is the sheer vastness that gives Australia – and its population – such immense character.

Best time to visit
Any time: when it's cold down south it's warm up north

Top things to see
- A concert, dance or theatrical performance at the country's most recognisable icon, the Sydney Opera House
- The red hues of Uluru, an awe-inspiring natural monolith that is both ancient and sacred
- Broome, where the desert meets the sea in contrasting aquamarines, rust-reds and pearl whites
- Provocative and engaging art at the world-class MONA, an eclectic museum housed in an underground lair in Hobart
- Paradisiacal beaches and an astounding underwater world as you island-hop around the Whitsundays on the Great Barrier Reef

Top things to do
- Discover Aboriginal culture, rock art and biological diversity at majestic Kakadu National Park
- Wind your way beside wild waters and craggy rock formations then whip inland through rainforests driving the Great Ocean Road
- Partner world-class wine and cosmopolitan dining at vineyard restaurants in the Barossa Valley or Margaret River
- Spy 'gentle giant' whale sharks at the World Heritage–listed Ningaloo Reef
- Kayak around the sheltered waters of Tasmania's Freycinet Peninsula

Getting under the skin
Read: *Cloudstreet*, Tim Winton's fascinating novel that chronicles the lives of two families thrown together in post-WWII Perth
Listen: to the authentic and insightful music of Aussie singer-songwriter Paul Kelly; or the soothing Aboriginal voice of Geoffrey Gurrumul Yunupingu on his album *Gurrumul*
Watch: *The Dry*, a murder mystery set against the challenges of drought faced by rural Australia
Eat: Sydney rock oysters; lightly-seared kangaroo meat; indigenous bushtucker ingredients

Drink: excellent craft beer at urban taprooms around every Australian city and suburb

In a word
G'day mate!

Celebrate this
Launched in 2020, Welcome to Country is a website showcasing the increasing range of indigenous travel and tourism experiences offered by Aboriginal groups and Torres Strait Islanders.

Random fact
Great Australian inventions include the bionic ear, the black box flight recorder and the wine cask.

1. Cradle Mountain-Lake St Clair National Park is in the heart of Tasmania

2. Bronte Baths ocean swimming pool is one of Sydney's finest; be here at 5am to watch the sun rise over the ocean

3. A guide from Kuku Yalanji tribe sitting beside Mossman River in Far North Queensland

Best time to visit
Year-round

Top things to see
- Vienna's *Christkindlmarkt* (Christmas market) with a mug of mulled wine in one hand and a bag of hot chestnuts in the other
- The opulent state apartments, prancing Lipizzaner stallions and jewels the size of golf balls in Vienna's palatial Hofburg
- Salzburg, the city where music, art and architecture achieve Baroque perfection
- Innsbruck's medieval heart, dwarfed by majestic snowcapped peaks
- Eisriesenwelt, the world's largest accessible ice caves deep in the heart of the mountains
- The Gothic masterpiece that is the Stephansdom, piercing Vienna's sky

Top things to do
- Revel in Vienna's extraordinary cultural offerings, rounded off with a performance at its celebrated opera house
- Book ahead for the best of the concerts and events at summer's Salzburg Festival
- Road-trip through the Hohe Tauern National Park along the scenic Grossglockner Road
- Dip into one glassy-blue lake after another in the Salzkammergut Lake District
- Hurl yourself on skis down the spectacular Harakiri – Austria's steepest slope – in Mayrhofen, or hobnob with the jet-set ski crowd in upmarket Lech

Getting under the skin
Read: *The Piano Teacher* (also a film) by 2004 Nobel laureate Elfriede Jelinek, to acquaint yourself with one of the most provocative Austrian writers
Listen: to Mozart, Haydn and Schubert
Watch: *The Sound of Music* for unbeatable songs and scenic views of Salzburg and its surrounding countryside
Eat: a bowl of soup speckled with *Knödel* (dumplings), followed by a *Wiener Schnitzel* (breaded veal or pork escalope) and sweet *Salzburger Nockerl* (fluffy soufflé)
Drink: *Sturm* (semifermented Heuriger wine) in autumn; *Glühwein* (hot spiced red wine) in winter; and coffee in a *kaffeehaus* any time of year

In a word
Grüss gott (Hello)

Celebrate this
If it had only produced Mozart, Austria would have had enough cause to celebrate its musical heritage, but Haydn, Schubert, Strauss and more help place the country in the top tier of cultural powerhouses.

Random fact
Vienna is the largest wine-growing city in the world.

1. Baroque architecture in medieval Innsbruck

2. The Grossglockner is the highest mountain in Austria

3. A snowy Salzburg overlooked by Hohensalzburg Fortress

A CAPITAL VIENNA // POPULATION 8.86 MILLION // AREA 83,871 SQ KM // OFFICIAL LANGUAGE GERMAN

Austria

Austria is sometimes like an outdoor film set, sometimes like a lavish and decadent Viennese ball where old-society dames waltz, gallop and polka with new-millennium drag queens. Be it the jewel-box Habsburg palaces and coffee houses of Vienna, the Baroque brilliance of Salzburg or the peaks that claw at the snowline above Innsbruck, the backdrop for travel in this landlocked Alpine country is phenomenal. Natural landscapes – giddy mountain vistas, dazzling glaciers and deep ravines – are as artful as a Mozart masterpiece or that romantic dance Strauss taught the world. Whether it's arts and culture you're into or scaling mountain peaks, in Austria you'll have a ball.

1. Learn about weaving techniques at the Carpet Museum in Baku

2. Deep in the Caucasus mountains near Russia, at an altitude of 2335m, Xınalıq is one of the highest villages in Europe

3. Past and future in the form of Baku's Old City and the Flame Towers

A   **CAPITAL** BAKU // **POPULATION** 10.2 MILLION // **AREA** 86,600 SQ KM // **OFFICIAL LANGUAGE** AZERI

Azerbaijan

It may not quite be paradise, but legend has it that Azerbaijan was the site of the Garden of Eden. Tucked beneath the Caucasus Mountains and hugging the Caspian Sea, it is where Central Asia blurs into Europe, with a mélange of Turkic, Persian and Russian influences contributing to its fabric. Azerbaijan offers dramatic and untouched mountain vistas, the gritty reality of Caspian oil rigs, ancient Zoroastrian temples, Bronze-Age petroglyphs and medieval caravanserais that have their historic roots in Central Asian trade. Like their Turkish brethren, Azerbaijanis are a passionate people who enjoy the timeless pleasures of a glass of *çay* (tea) with friends over a spirited game of *nard* (backgammon).

Best time to visit
May to June and September to November for pleasant weather and fewer crowds

Top things to see
- Baku's futuristic architecture, including the iconic 'flickering' Flame Towers and the Zaha Hadid–designed Heydar Aliyev Center
- Charming Şəki, Azerbaijan's loveliest town, whose traditional centre was declared a Unesco World Heritage Site in 2019
- The coppersmiths of Lahic, hammering away to create the town's renowned artwork

Top things to do
- Get lost in the atmosphere of Baku's Old City, wandering the narrow alleyways around the Palace of the Shirvanshahs
- Take a horse ride into the Caucasus mountains, following ancient trails with a local guide
- Marvel at ancient petroglyphs and the sputtering mud volcanoes of starkly beautiful Qobustan

Getting under the skin
Read: *Ali and Nino*, Kurban Said's epic novel about a doomed love affair between an Azerbaijani Muslim and a Georgian Christian, set in early-20th-century Baku
Listen: to *muğam*, Azerbaijan's very own musical idiom, a traditional minstrel art form that's emotional and spine tingling
Watch: Ilgar Najaf's *The Pomegranate Orchard*, an Azerbaijani entry for Best Foreign Film Oscar in 2018
Eat: flame-grilled *shashlyk* (lamb kebab)
Drink: *çay* (tea) at a traditional teahouse

In a word
Salam (Hello)

Celebrate this
Azerbaijan has hosted high-profile sporting events, including annual Formula One races and UEFA matches, though critics accuse the government of using them to 'sportwash' its human rights abuses.

Random fact
'Layla', Eric Clapton's classic rock song, was inspired by the Azeri epic poem *Layla and Majnun*.

1. Pastel-pink Government House in Nassau

2. The Insta-famous pigs of Big Major Cay in the Exumas; visitors should behave responsibly

3. Inagua National Park hosts the largest breeding colony of West Indian flamingos in the world (the national bird of the Bahamas)

B **CAPITAL** NASSAU // **POPULATION** 319,031 // **AREA** 13,940 SQ KM // **OFFICIAL LANGUAGE** ENGLISH

Bahamas

If you visited an island a day, you'd have over eight years of perfect adventure in the Bahamas. Some 3100 islands and islets dot the archipelago. Some are no bigger than a limestone spit poking above the swells; others are fully fledged paradise islands, with swaying palms and blond-sand beaches. The Bahamas has a reputation as a luxe tourist destination, but it's easy to escape this commercial vibe, particularly on the appropriately named Out Islands, or under the tropical waters, which you share with turtles and reef sharks.

Best time to visit
Year-round: December to February to escape northern cold, June to August for full-on tropical heat and humidity

Top things to see
- Nassau's pirate museums and nightlife
- Inagua National Park, roosting spot for pink flamingos
- Harbour Island, where the sands – not the birds – are pink
- Long Island, with pastel-hued houses and over 120km of blissfully empty beaches
- Cat Island, a refuge for traditional Bahamian culture and the best place to hear local 'rake and scrape' music

Top things to do
- Kayak among dozens of tiny islands (cays) in the Exumas
- Island-hop by mail boat
- Savour grilled conch at a beach shack on Grand Bahama
- Take the plunge at Small Hope Bay, with shark and reef dives in the lagoon
- Explore a shipwreck (there's one or two for every island)

Getting under the skin
Read: *Carnival of Love* by Ernestia Fraser, which follows a Bahamian family while centering on island culture
Listen: to Tony Mackay, alias Exuma, from Cat Island; his classic Caribbean-themed songs include 'The Obeah Man'
Watch: *Womanish Ways: Freedom, Human Rights & Democracy*, a film about the women's suffrage movement in the Bahamas by writer and film maker Marion Bethel
Eat: conch (a mollusc served pounded or minced; marinated or grilled; or even raw as ceviche)
Drink: Kalik (a light, sweet lager); or a goombay smash, a dangerously easy-to-quaff rum punch

In a word
What da wybe is? (What's going on?)

Celebrate this
Junkanoo – it's a music style as well as the country's national festival. Each year on Boxing Day, New Year's Day and again for one day in the summer, teams compete to have Nassau's best costumes, music, dancing and floats.

Random fact
Many Bahamians practice *obeah*, a ritualistic form of magic with deep African roots.

CAPITAL MANAMA // **POPULATION** 1.5 MILLION // **AREA** 760 SQ KM // **OFFICIAL LANGUAGE** ARABIC

Bahrain

The smallest of all Arab countries, Bahrain is – its Formula One Grand Prix aside – the least glitzy of the Gulf emirates. While the country has embraced the modern with Manama's seafront swoop of contemporary architecture, the past remains a presence in the echoes of the ancient Dilmun civilisation and in the former pearl-fishing district of Muharraq. Tensions between Bahrain's Shia Muslim community and its Sunni Muslim ruling family came to the fore during 2011's civil unrest and following crackdown. They continue to simmer a decade on due to ongoing repression.

Best time to visit
October to April

Top things to see
- Five thousand years of history under one roof at Bahrain National Museum
- The Portuguese-era Bahrain Fort, built atop a settlement mound dating back to the Bronze Age Dilmun civilisation
- Unesco World Heritage–listed Muharraq Island with its cache of historic buildings, souqs and heritage museums
- Manama Souq, for a traditional shopping experience
- Al-Areen Wildlife Park and Reserve, where you can see 240 bird species and Arabian oryx

Top things to do
- Scuba dive for pearls then visit the Museum of Pearl Diving to learn all about Bahrain's pearling heritage
- Watch Bahrain put on a show for the Formula One Grand Prix, usually held in April
- Tour the elegant interior of Manama's Al-Fatih Mosque
- See a performance at Bahrain National Theatre, Manama's seafront architectural showpiece
- Take a mountain bike or trekking tour on Hawar Island to spot Arabian oryx and Nubian ibex

Getting under the skin
Read: *The Randomist,* an essay collection by Ali Al Saeed; or Geoffrey Bibby's *Looking for Dilmun*, a story of archaeological treasure-hunting and a window onto 1950s and '60s Bahrain
Listen: to *Desert Beat* by Hashim al-Alawi
Watch: *Al-Hajiz* (The Barrier), *Za'er* (Visitor) or *A Bahraini Tale*, all directed by Bassam Al Thawadi – they're the only three films ever made in Bahrain
Eat: *muhammar* (honey-sweetened rice, spiked with cardamom and saffron) served with grilled Gulf fish; and *khanfaroosh* (cardamom and saffron fried cakes)
Drink: cardamom-infused Arabic coffee

In a word
Al-hamdu lillah (Thanks to God)

Celebrate this
In 2019, Bahrain's over 10,000 Dilmun civilisation burial mounds, spanning 21 archaeological sites, and dating back to the 2nd-century BCE, gained Unesco World Heritage listing.

Random fact
Bahrain originally consisted of 33 islands, but that number is increasing, as is the length of its coastline, as more and more land is (somewhat controversially) reclaimed from the sea.

1. Bahrain's World Trade Center stands 240m high

2. Winter is the best time to explore the desert dunes, when temperatures are lower

3. Stained glass at the Beit al-Quran Museum in Manama

B CAPITAL DHAKA // POPULATION 162.7 MILLION // AREA 143,998 SQ KM // OFFICIAL LANGUAGE BENGALI

Bangladesh

Bangladesh is the South Asia most travellers miss in the headlong rush to reach India, Nepal and Sri Lanka. In this fascinating Asian nation, rivers function as main roads, rickshaws are mobile movie posters, hills are painted emerald green by tea plantations, and beaches stretch beyond the horizon, unseen by all but a tiny contingent of hard-core travellers who relish the buzz of stepping off the tourist trail. Before 1947, Bangladesh was the eastern half of Indian Bengal; then from 1947 to 1971 it was East Pakistan. Today, this famously crowded Muslim nation is struggling to forge its own identity in the face of challenges ranging from religious extremism and poverty to floods, cyclones and climate change. But the waterlogged landscape is also part of Bangladesh's charm – like stepping into a vanished monsoon summer from an earlier age.

Best time to visit
October to February, avoiding the torrential monsoon

Top things to see
- The unbelievable crowds of boats and people at the Sadarghat docks in Dhaka
- The colours of tea-pickers' saris against the green of tea plantations near Srimangal and Sylhet
- A Royal Bengal tiger (if you're lucky) stalking the waterways of Sundarbans National Park
- Buddhist stupas and Adivasi tribal culture in the Chittagong Hill Tracts
- Tropical waves breaking on the white sands of tiny St Martin's Island

Top things to do
- Experience the surreal phenomenon of a rickshaw traffic jam in Dhaka
- Ride the Rocket – the rickety paddle-wheel ferry that trundles from Dhaka to Morrelganj
- Feel the liberation of wearing a *lungi* (sarong)
- Drop in on the Buddhist tribal villages dotted around remote Kaptai Lake
- Stroll barefoot along the world's longest beach – now a surprising surfing hub – near Cox's Bazar

Getting under the skin
Read: Tahmina Anam's *A Golden Age*, the story of the Bangladesh War of Independence through the eyes of one family
Listen: to the rousing poems of Bangladesh's national poet Kari Nazrul Islam.
Watch: Satyajit Ray's Bengali classic *Apu Trilogy*; or Rubaiyat Hossain's award-winning drama about textile workers, *Made in Bangladesh*
Eat: *ilish macher paturi* (a classic Bengali dish of hilsa fish steamed inside banana leaves)
Drink: *sharbat* (chilled yoghurt mixed with chilli, coriander, cumin and mint) – the perfect accompaniment to a fiery curry

In a word
Tik aache (No problem)

Celebrate this
As head of the Bangladesh Centre for Worker Solidarity, former garment worker Kalpona Akter is leading the push for better working conditions for the 3.5 million Bangladeshi workers who toil making clothes for international fashion brands.

Random fact
The national game of Bangladesh is *kabaddi*, a group version of tag where players must evade the opposing team while holding a single breath of air.

1. A floating market in Bangladesh where guava is the commodity of choice

2. Tea, being harvested here at Srimangal, is the country's second-largest export crop

3. There are about 800,000 rickshaw-pullers in Dhaka

1. You can lead a horse to water on Pebbles Beach, if you wish to wash it

2. For those without a yacht, there are plenty of charter businesses in Barbados

3. St Nicholas Abbey was built as a sugarcane plantation house and is now a museum and rum distillery

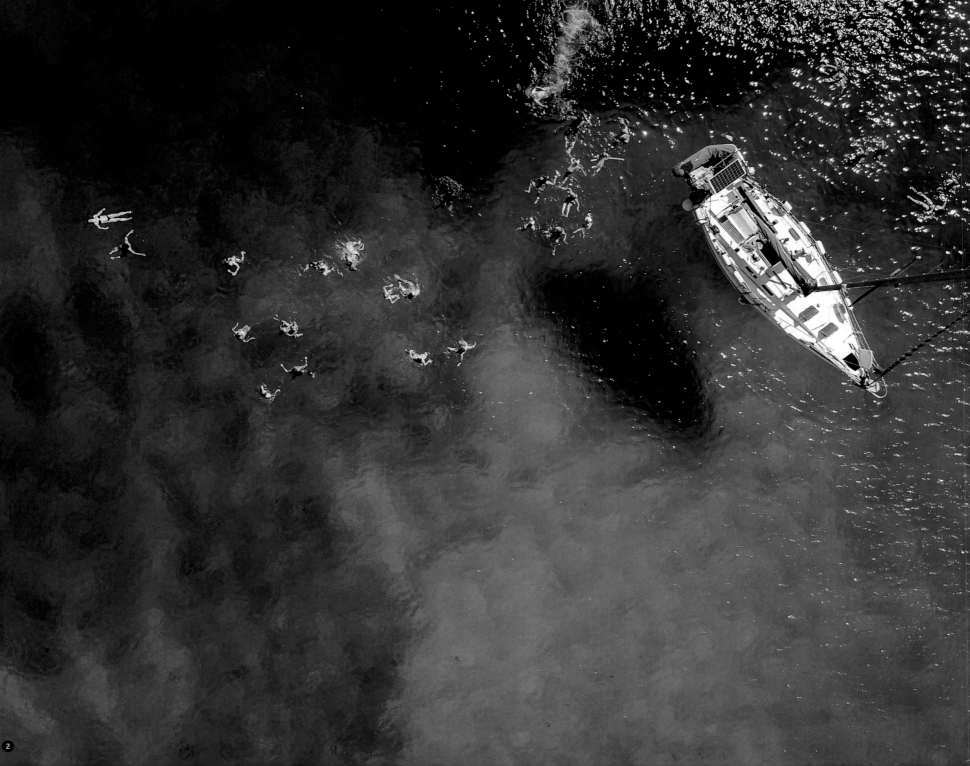

CAPITAL BRIDGETOWN // POPULATION 288,725 // AREA 431 SQ KM // OFFICIAL LANGUAGE ENGLISH

Barbados

You can drive almost all day on Barbados and never see a beach. This pork-chop-shaped island has a deep interior traversed by narrow, winding roads that pass through isolated hamlets, old plantations and tiny villages. Beach bums take heart – all roads eventually lead to the coast, which is ringed with strands of white sand, each with a distinct personality. Find wave-tossed surfer havens in the east; palm-fringed bays with resorts great and small in the west.

Best time to visit
February to May

Top things to see
- Barclay's Park, with a seashore ripe for beachcombing
- St Nicholas Abbey, which combines the beauty of a plantation estate with horrific revelations about slavery
- Hunte's Gardens, a jungle oasis of flowers, trees and birds set in a tropical gulley
- Lavishly restored George Washington House, the US president's home for two months in 1751

Top things to do
- Cheer madly for your team at a cricket match
- Surf the legendary Soup Bowl on the east coast
- Get in the groove of Bridgetown, the island's capital and home to dockside cafes, shops and thriving markets
- Explore the tiny coastal road to Fustic, where ages-old fishing enclaves snooze amid riots of flowers

Getting under the skin
Read: *How the One-Armed Sister Sweeps Her House* by Cherie Jones – a powerful novel about four people trying to escape a legacy of violence in beautiful Barbados
Listen: to calypso artist the Mighty Gabby; and soca artist Rupee
Watch: *A Caribbean Dream* – a retelling of Shakespeare's *A Midsummer Night's Dream* set on the backdrop of Carnival in Barbados
Eat: a range of dishes with African roots adapted to local produce, such as *cou-cou* (a creamy cornmeal and okra mash, often served with saltfish) and *jug-jug* (a mix of cornmeal, peas and salted meat)
Drink: world-renowned Mount Gay rum at a floating club

In a word
Workin' up (Dancing)

Celebrate this
Barbados celebrates music of all kinds. Beyond Bajan star Rihanna, there is calypso, soca, rapso and dancehall and venues all over the island in which to dance.

Random fact
Barbados boasts more world-class cricket players on a per capita basis than any other nation.

Best time to visit
May to September

Top things to see
- Minsk, with its bombastic communist architecture, top-notch ballet and opera, and unofficial arts district of Vul Kastrychnitskaya
- The impressively restored Mir Castle, which looks like something out of a fairy tale
- The extraordinary Brest Fortress, a WWII memorial that commemorates Soviet resistance against the German invasion in the spot where Operation Barbarossa was unleashed
- Dudutki, a reconstructed 19th-century Belarusian village with craft-making exhibits, horse riding and local delicacies, including *samagon* (moonshine)

Top things to do
- Observe Europe's only surviving wild bison (the *zoobr*) at the Belavezhskaya Pushcha National Park, the oldest wildlife refuge in Europe
- Follow the masochistic recipe for good health at a Belarusian bathhouse: sit in a steam room, beat yourself with damp twigs, dunk yourself in icy water, repeat
- Wander the old neighbourhoods of Vitsebsk, childhood home of famous painter Marc Chagall, and visit his museum
- Explore the vast swathe of marshes, swampland and floodplains of the relatively untouched Pripyatsky National Park

Getting under the skin
Read: *Voices from Chernobyl*, which records the effects of the disaster on the people of the region, by Nobel prize–winner Svetlana Alexeivich
Listen: to the vaguely apocalyptic improvisations of Knyaz Myshkin, a Minsk-based band
Watch: Darya Zhuk's debut *Crystal Swan*, a story about an aspiring DJ who dreams of moving to the USA
Eat: *solyanka* (meat, potato and pickled vegetable soup); and *draniki* (potato pancakes)
Drink: *kvass* (an elixir made of malt, flour, sugar, mint and fruit)

In a word
Vitayu (Hello)

Celebrate this
The underground theatre group, Belarus Free Theatre, has been producing plays and opposing all kinds of artistic censorship and totalitarianism since 2005.

Random fact
Belavezhskaya Pushcha National Park has some 1300 sq km of primeval forest; half of the park lies in Poland.

1. The heroic monument known as 'Courage' at Brest Fortress

2. About 300 European bison live in Belavezhskaya Pushcha National Park

3. The giant Gates of Minsk on Railway Square

CAPITAL MINSK // **POPULATION** 9.4 MILLION // **AREA** 207,600 SQ KM // **OFFICIAL LANGUAGES** BELARUSIAN & RUSSIAN

Belarus

Having suddenly emerged on travellers' radars thanks to relaxed visa requirements, Belarus is at first glance a country that has never entirely left the USSR and doesn't tolerate dissent. But while this may be a land of communist-style architecture, state-run media and a centralised economy, you'll also encounter some excellent art galleries, hip bars and restaurants. Leave the archetypally Soviet capital Minsk and you'll find yourself in a bucolic land of forests, rivers, fairy-tale castles and Slavic folk traditions. Indeed, Belarusians are people who still relish life's simple pleasures: weekends spent at the *dacha* (summer country house), mushroom-picking in the woods or steaming away at the *banya* (bathhouse).

Best time to visit
May to September

Top things to see
- Brussels' guildhalls of medieval craftsmen on Grand Place
- The sobering war memorials of Flanders Fields and Waterloo Battlefield
- Galeries St Hubert, the grande dame of 19th-century shopping arcades
- Medieval Bruges with its cobblestone lanes, romantic canals and historic *begijnhof* (inner courtyards)
- Belgium's largest Gothic cathedral, Onze-LieveVrouwekathedraal, in Antwerp, which houses four paintings by Rubens

Top things to do
- Admire the historical and cultural riches at Brussels' Musée Art & Histoire
- Expand your waistline by indulging in pralines and truffles by prized chocolate-maker Pierre Marcolini
- Shop for fashion by local designers, fine-dine and dance until dawn in Antwerp
- Explore Ghent's squares and markets by bike
- Peruse lengthy beer menus and sample abbey-brewed Trappist dark ales in pubs across the country

Getting under the skin
Read: *The Lady and the Unicorn* by Tracy Chevalier for a glimpse of Brussels' medieval master tapestry makers

Listen: to harmonica virtuoso Toots Thielemans blast the best of Belgian jazz with *Hard to Say Goodbye – The Very Best of Toots Thielemans*

Watch: *Le Silence de Lorna* (Lorna's Silence) – the grim tale of an Albanian girl growing up in Belgium – by Belgian film-making brothers Jean-Pierre and Luc Dardenne

Eat: a steaming cauldron of mussels cooked in white wine with a mountain of *frieten/frites* (fries or chips)

Drink: Trappist beer brewed by monks in Rochefort

In a word
Proost/Santé (Cheers in Dutch/French)

Celebrate this
Trappist, Abbey, lambic, wheat and even a champagne beer – Belgium is a small country with a big appetite for beer in all its varieties.

Random fact
Belgium's linguistic divide was formalised in law in 1962 when the official line was drawn between Dutch-speaking Flanders and French-speaking Wallonia.

1. The name of this Brussels pub, dating from 1928, translates as 'to the sudden death'

2. The Tyne Cot Commonwealth WWI Graves Cemetery and Memorial to the Missing is near Ypres

3. The view from the belfry over Bruges

B

CAPITAL BRUSSELS // POPULATION 11.72 MILLION // AREA 30,528 SQ KM // OFFICIAL LANGUAGES DUTCH, FRENCH & GERMAN

Belgium

There are few places that feel more in the heart of Europe than Belgium, a multilingual country smack-bang in the middle of Western Europe that Dutch-speaking Flemish in the north call België, and French-speaking Walloons in the south call La Belgique. Indeed, it's this north–south cultural and linguistic divide that makes this intimate country so unexpectedly fascinating. Amble in the shade of art nouveau architecture in its capital, Brussels; give a nod to the EU headquarters; surrender to the sweet seduction of the finest chocolate; immerse yourself in Antwerp's vibrant fashion scene and cool clubs; and down a fine local beer or three.

Best time to visit
January to May (when there is less rain)

Top things to see
- The *cayes* (islands) strung out along the 1000km-long Meso-American Barrier Reef
- Black howler monkeys at the Community Baboon Sanctuary near Bermudian Landing
- Caracol, a vast Mayan city once home to over 100,000 people, now tucked into matted jungle
- Wild dancing and wilder costumes at September's Belize Carnival in Belize City
- Orderly and cosmopolitan Belmopan, the smallest capital city in the Americas

Top things to do
- Groove to live punta rock, which fuses tribal Garífuna beats, soca and reggae
- Ride a speedboat from Belize City to Caye Caulker, for the full Caribbean island experience
- Explore the narrow passages of Barton Creek Cave, an ancient sacred site for Mayan burials
- Dive the submarine canyons, cuts and blue holes of Caye Ambergris and the Northern Cayes
- Look for the 5ft-tall jabiru stork, the tallest bird in the Americas, at Crooked Tree Wildlife Sanctuary

Getting under the skin
Read: *The Last Flight of the Scarlet Macaw*, Bruce Barcott's account of one activist making a difference in Belize
Listen: to the Garífuna rhythms and culture of Andy Palacio's *Wátina*
Watch: Harrison Ford face jungle fever in *The Mosquito Coast*, with the Belize interior standing in for neighbouring Honduras
Eat: the Belizean staples of rice and beans, perfect with a dash of Marie Sharp's famous hot sauce
Drink: the sweet water of green coconuts split open by machete, or traditional cashew wine at the Cashew Festival in Crooked Tree

In a word
Arright? (You all right? – the ubiquitous greeting)

Celebrate this
Founded in 1986, Cockscomb Basin Wildlife Sanctuary was the first ever protected area dedicated to the jaguar. It now stretches over some 500 sq km of forests and foothills and is one of the most important sites for the animal's preservation in the world.

Random fact
Released red lionfish, native to the Pacific, are thriving in the Caribbean, where they threaten local species. The government is encouraging restaurants to serve up the invader, which tastes like snapper.

1. The Mayan settlement of Altun Ha was an important trading town; conservation of the ruins began in 2000

2. There are more than 600 bird species in Belize, including this keel-billed toucan

3. Portrait of a Mayan man from San Antonio

4. Tobacco Caye is a tiny island 12 miles off the coast of Belize

B CAPITAL BELMOPAN // POPULATION 383,000 // AREA 22,966 SQ KM // OFFICIAL LANGUAGE ENGLISH

Belize

The Spanish weren't the only people to carve out an empire in Central America. Tiny, jungle-cloaked Belize was the site of Britain's brief colonial experiment in this tropical isthmus. The legacy lives on in Belize's distinctive Afro-Caribbean culture, and in the lilting English spoken by most residents. Compact Belmopan has been the capital since 1970, although bustling, chaotic, Belize City remains the main commercial centre. Inland are dense rainforests dotted with Mayan ruins, and offshore is a shimmering necklace of tropical islands, following the line of the world's second-largest coral reef. Most visitors arrive by cruise ship, drop in for the day and move on, leaving the treasures inland for more committed travellers. For the adventurous, Belize is a handy stepping stone on the overland trail to Guatemala, El Salvador and beyond.

Best time to visit
November to February (the dry season)

Top things to see
- The enduring monuments to the kings of Dahomey in Abomey
- The Point of No Return memorial for countless enslaved African people in Ouidah
- Grand Popo, a palm-fringed beach town on Benin's Gulf of Guinea coast
- Contemporary African art at Ouidah's remarkable Zinzou Foundation Museum
- Dassa Zoumé, Benin's most striking terrain with rocky outcrops and lush hills

Top things to do
- Safari in Parc National de la Pendjari, home to herds of elephants and prides of rare West African lions
- Trek through the rugged northern Atakora region with a local ecotourism association
- Spend a night sleeping above the waters in the stilted village of Ganvié
- Learn traditional fishing techniques at Lake Ahémé
- Dance deep into the night on 10 January, Benin's Voodoo Day

Getting under the skin
Read: *Dawn to Dusk: Folktales from Benin*, a collection of oral history that speaks of Edo culture and tradition; while Annie Caulfield's taxi journey around Benin is wittily told in *Show Me the Magic*
Listen: to anything by Angelique Kidjo, Gangbe Brass Band or Orchestre Poly-Rythmo
Watch: *In Search of Voodoo: Roots to Heaven,* an insightful look into the often-misrepresented religion
Eat: *igname pilé* (pounded yam served with vegetables and meat) and *gombo* (okra)
Drink: *tchouukoutou*, a traditional millet-based brew

In a word
Neh àh dèh gbò? (How are you?)

Celebrate this
Singer, actor, activist and Unicef Goodwill Ambassador Angelique Kidjo: she continues to make waves for all the right reasons.

Random fact
Despite Benin being the birthplace of voodoo, the religion was not formally recognised here until February 1996.

1. Wild kob antelope in Parc National de la Pendjari

2. A dancer at the Ouidah Voodoo Festival – some 40% of Benin's population follow voodoo, which holds that all of creation is divine

3. Benin has about 75 miles of Atlantic coast

B CAPITAL PORTO NOVO // POPULATION 12.9 MILLION // AREA 112,622 SQ KM // OFFICIAL LANGUAGE FRENCH

Benin

Once divided and fought over by many powerful West African kings, the petite nation of Benin is loaded with extraordinary vestiges of historical significance. Palaces speak of wealth, power and bloodshed, while forts and ports provide a sobering insight into the nation's prominent role in the centuries-long trans-Atlantic trade in human beings. Some evocative sights and events speak as much to the past as the present, such as those involving voodoo –Benin is the birthplace of this often-misunderstood religion, and it's still widely practised today. A beacon of stability and home to sustainable ecotourism projects and some of Africa's rarest wildlife, Benin works hard to be a success. Pick through clamorous markets, lounge on tropical beaches and watch wildlife – you'll find it easy to fall in love with Benin.

Best time to visit
March to May, October to November

Top things to see
- Markets, museums, artisans' workshops and surprisingly vibrant nightlife in the capital Thimphu
- Taktshang Goemba (Tiger's Nest Monastery), balancing on a cliff wall and reputedly held in place by the hairs of angels
- Gorgeous Punakha Dzong, straddling the junction of the Mo (Mother) and Po (Father) Rivers
- Serene sacred sites and rhododendron forests in the Bumthang valley
- Terrific Trongsa, whose enormous *dzong* (monastery fortress) once guarded the only route crossing the country

Top things to do
- Attend morning prayers and sip salted butter tea with the monks at a Bhutanese monastery
- Trek along ancient Himalayan trade routes to the sacred peak of Jhomolhari
- Marvel at the accuracy, banter and bravado of an archery tournament in Thimphu
- Spot rare black-necked cranes in the glacier-carved Phobjikha Valley
- Be amazed by the colours and costumes at a traditional *tsechu* (Buddhist dance festival)

Getting under the skin
Read: expat insights in *Beyond the Sky and Earth* by Jamie Zeppa; or *Buttertea at Sunrise: A Year in the Bhutan Himalaya* by Britta Das
Listen: to *Endless Songs from Bhutan*, a collection of Bhutanese folks songs by Jigme Drukpa
Watch: *Travellers and Magicians*, directed by reincarnated lama Khyentse Norbu; or Pawo Choyning Dorji's *Lunana: A Yak in the Classroom*, exploring life at a remote Bhutanese school
Eat: mouth-melting *ema datse* (green chillies and cheese); or, if you dare, *aphrodisiac cordyceps*, a parasitic fungus that turns insects into zombies!
Drink: the craft beers of the Namgyal Artisanal Brewery; or *ara*, Bhutan's home-grown firewater

In a word
Kuzuzangbo la (Hello)

Celebrate this
The third Druk Gyalpo (Dragon King) of Bhutan, Jigme Singye Wangchuck, is credited with introducing Bhutan's policy of measuring development in terms of Gross National Happiness, a philosophy that still guides government decisions.

Random fact
The king of Bhutan, Jigme Khesar Namgyel Wangchuck, is an enthusiastic mountain biker; he's often spotted barrelling down the trails around the capital.

1. Guru Rinpoche is said to have to flown to Taktshang Goemba on the back of a tigress

2. Masked dances are part of Bhutan's spiritual and cultural traditions

3. Massive Paro Dzong is home to 200 monks

KELLY CHENG TRAVEL PHOTOGRAPHY | GETTY IMAGES // DYLAN HASKIN | SHUTTERSTOCK // JONATHAN GREGSON | LONELY PLANET

CAPITAL THIMPHU // **POPULATION** 782,318 // **AREA** 38,394 SQ KM // **OFFICIAL LANGUAGE** DZONGKHA

Bhutan

The last Himalayan kingdom, Bhutan holds an almost mythical status among travellers for its beguiling combination of mystery and inaccessibility. Those who can afford the US$200-250 daily fee can trek through valleys scented by blue pines, to fortress-like monasteries that have changed only superficially since Padmasambhava first transported Buddhism across the Himalaya. Bhutan's full hand of mountains, monasteries and magic is only part of the story; defying preconceptions, this is also a nation of smart phones, fast broadband and espresso bars. Bhutan is modernising, but on its own terms, and locals are proud that their government prioritises Gross National Happiness over Gross National Product. It's an enigmatic package that is not really like anywhere else on Earth.

CAPITAL LA PAZ (ADMINISTRATIVE) & SUCRE (CONSTITUTIONAL) // **POPULATION** 11.6 MILLION //
AREA 1,098,581 SQ KM // **OFFICIAL LANGUAGES** SPANISH, QUECHUA, AYMARA & OTHERS

Bolivia

Bolivia is a country of epic landscapes. Here, 6000m Andean peaks meet high-altitude deserts, surreal salt flats, Amazonian rainforest and the savannahs of the Pantanal. Its people are almost as varied, with 37 official languages and indigenous groups such as the Quechua and Aymara accounting for over 50% of the population. Thanks in part to Bolivia's relative isolation, traditions are rarely far away, and bright textiles and wild festivals help make the country one of the continent's most distinctive. Yet this is also a place of scarcity, with one of the world's lowest population densities. Bolivia's GDP is the lowest in South America, despite extensive mineral deposits, although this century has seen impressive economic growth.

Best time to visit
May to October

Top things to see
- The breath-sapping city of Potosí and its bleak mining legacy
- The Salar de Uyuni, where blindingly white salt deserts complete with bubbling geysers and aquamarine lagoons
- The dizzying city of La Paz, set deep in a canyon and fringed by snow-covered mountains
- The astounding range of wildlife in the rainforest, mountains and grasslands of Parque Nacional Madidi, near Rurrenabaque

Top things to do
- Trek through the giddying heights of the Cordillera Real along ancient Inca routes
- Feel the rush on a cycle ride along what was once dubbed 'the world's most dangerous road'
- Boat it out to the tradition-steeped Isla del Sol, legendary island home of the Inca on the world's highest navigable lake, Lake Titicaca
- Party the night away at February's Carnaval de Oruro, the nation's biggest, brightest festival

Getting under the skin
Read: *The Fat Man from La Paz: Contemporary Fiction from Bolivia*, edited by Rosario Santos, which contains 20 short stories illustrating life in Bolivia
Listen: to charango master Celestino Campos or famous Andean folk trio Los Kjarkas
Watch: *The Devil's Miner*, Kief Davidson and Richard Ladkani's award-winning documentary following two brothers working in one of South America's most dangerous mines
Eat: *salteña* (a meat and vegetable empanada); *surubi* (catfish)
Drink: the mildly alcoholic *chicha cochabambina*, made from fermented corn; *mate de coca* (coca leaf tea), which helps ease altitude sickness

In a word
Que tal? (How are you?)

Celebrate this
Evo Morales, Bolivia's first indigenous president (2006–2019), left office with questions over environmental policies and electoral fraud. Yet in bringing indigenous Bolivians into senior roles and slashing poverty, he gave the dispossessed a voice.

Random fact
Many of the riches of Spain flowed from Cerro Rico above Potosí, where an estimated 45,000 tons of pure silver was mined between 1550 and 1780.

1. La Paz's Mi Teleférico cable car system is said to be the world's longest

2. High on Bolivia's altiplano, flamingos congregate on Laguna de Canapa

3. Traditional costumes at Fiesta de la Virgen de la Candelaria in Copacabana

Best time to visit
April to September

Top things to see
- Daredevils plunging from Mostar's bridge into the racing green waters below
- Sarajevo, the 'Jerusalem of the Balkans', a vibrant city with cultural events and nightlife aplenty
- The castle and Ottoman-era architecture of Travnik, once the seat of Bosnia's governors
- The medieval Bosnian capital of Jajce, cut by the Pliva waterfall and defended by imposing gates
- The Unesco-listed bridge in Višegrad, which inspired the Nobel Prize–winning novel *Bridge on the Drina* by Ivo Andrić

Top things to do
- Wander the cobbled alleys, coppersmith workshops and artisanal stalls of Baščaršija in Sarajevo
- Join the pilgrims waiting for an apparition of Mary in Međugorje, or nod at Sufi tombs in the Tekija (dervish lodge) in Blagaj
- Plummet down the Vrbas or Una rivers on a raft or in your own kayak
- Hike the Via Dinarica through Sutjeska National Park or hit bargain-value 1984 Olympic slopes on Jahorina and Bjelašnica mountains

Getting under the skin
Read: *Sarajevo Marlboro* by Miljenko Jergović, tales set during the siege of Sarajevo; or *Quiet Flows the Una* by Faruk Šehić, winner of the EU Prize for Literature
Listen: to *sevdah* music, Bosnia's folk music, an excursion in harmonious melancholy
Watch: Danis Tanović's *No Man's Land* or *Grbavica* by Jasmila Žbanić, both poignant reflections on the Bosnian war
Eat: *burek* (cylindrical filo pastry filled with minced meat); *sirnica* is the same pastry filled with cheese and *zeljanica* with spinach
Drink: the local dry red wine, Blatina; or *Bosanska kafa/kava* (Turkish coffee, served in a brass pot)

In a word
Živjeli (Cheers)

Celebrate this
The greatest treasure of the National Museum of B&H is the Sarajevo Haggadah, a 14th-century illuminated codex used during Passover. It survived the expulsion of Sephardic Jews from Spain, was hidden in a mosque during the Nazi occupation of Sarajevo in WWII, and saved from bombardment during the 1990s Sarajevo siege.

Random fact
Perućica, part of the Sutjeska National Park, is one of the last remaining primeval forests in Europe.

1. The Sutjeska National Park Monument commemorates fallen soldiers in the 1943 Battle of the Sutjeska

2. A shopping street of Sarajevo's Old Town

3. Pliva waterfall in the centre of Jajce

B | **CAPITAL** SARAJEVO // **POPULATION** 3.8 MILLION // **AREA** 51,197 SQ KM // **OFFICIAL LANGUAGES** BOSNIAN, CROATIAN & SERBIAN

Bosnia & Hercegovina

This Southeastern European crossroads is a place of charm and beauty, with plenty of outdoors adventures to match its cultural complexity. The variety of this 'heart-shaped land' is astounding: the cafe culture of Sarajevo, semi-ruined medieval fortresses, muezzins calling the faithful to prayer from Bosniak (Bosnian Muslim) minarets, austere architectural vestiges of Yugoslavia's communism, Orthodox monasteries clinging to hillsides – it can be hard to pin down the 'true' Bosnia and Hercegovina. Ultimately, it's beyond the cities that this deeply traditional country best expresses itself, through spreading vineyards, mountain villages and a relaxed pace of life.

B | **CAPITAL** GABORONE // **POPULATION** 2.3 MILLION // **AREA** 581,730 SQ KM // **OFFICIAL LANGUAGE** ENGLISH

Botswana

Diamonds are not forever, but in Botswana their legacy may just be. First discovered here by industrious termites (yes, really) in the late 1960s, the gems have allowed one of the world's poorest countries to transform itself and its economy. With billions of dollars of mining revenue being spent on healthcare, education and infrastructure, the path is being paved for a sustainable future. Key to this is the government's recognition that Botswana's most valuable riches may not be buried treasure, but rather its unparalleled wilderness and the iconic wildlife that calls it home. As such, vast tracts of the country are protected to provide a wealth of safari and tourism experiences.

Best time to visit
May to September (dry season) for classic safaris; November to April (wet season) for birds

Top things to see
- Some of the world's largest herds of elephants parading along the banks of the Chobe River
- The black-maned lions in the vast expanses of the Central Kalahari Game Reserve
- Thousands of zebra migrating into Makgadikgadi Pans National Park
- The sandstone landscape of the Tuli Block set alight by the setting sun
- San rock paintings that date back millennia, particularly in the remote Tsodilo Hills – where the natural gallery of rock formations is equally riveting

Top things to do
- Lie down on the Makgadikgadi Pan at sunset and watch the heavens appear above
- Explore the depths of the Okavango Delta's treasures in a *mokoro* (traditional canoe)
- Enjoy some of the planet's best wildlife watching on a dawn drive in the Moremi Wildlife Reserve
- Take some time to understand San culture in D'kar

Getting under the skin
Read: *Serowe: Village of the Rain Wind* by Bessie Head for an understanding of Tswana; *The No.1 Ladies' Detective Agency* by Alexander McCall Smith
Listen: to the Wizards, who blend hip hop with ragga and R&B; or anything by *kwasa kwasa* king Franco
Watch: *Into the Okavango*, a documentary following explorers on a four-month expedition down the Okavango River
Eat: *leputshe* (wild pumpkin) atop *bogobe* (sorghum porridge)
Drink: the stiff concoction from fermented marula fruit

In a word
Dumela (Hello, in Tswana) – extra marks if it's done with a 'special' handshake (place your left hand on your elbow while shaking

Celebrate this
Unity Dow: a long-time human rights activist who was the first woman to be appointed a high court judge in Botswana; she's also an acclaimed novelist of legal thrillers and became Minister of Foreign Affairs and International Cooperation.

Random fact
There are more elephants in Botswana than anywhere else in Africa.

1. Wildebeest race across the Makgadikgadi Pans National Park

2. Black-backed jackals greet each other in the Kalahari Desert

3. The San people are among the oldest inhabitants of southern Africa

1. Among the prey of the Pantanal's jaguars are large caimans

2. A Christmas procession in Salvador da Bahia, the heart of Brazil's Afro-Brazilian community

3. Brasília, a purpose-built capital city, was founded in 1960 and is packed with modern architecture by the likes of Oscar Niemeyer, such as the Supreme Federal Court

CAPITAL BRASÍLIA // **POPULATION** 211.7 MILLION // **AREA** 8,515,770 SQ KM // **OFFICIAL LANGUAGE** PORTUGUESE

Brazil

South America's largest and arguably most seductive country is accustomed to turning heads – and not merely for its breathtaking beaches and hedonistic festivals. This is Latin America's largest economy, and the host of both the Olympics and the football World Cup in the 2010s. It's not all been plain sailing, with political scandals and an economic slowdown in recent years, but Brazil is a major world player and a compelling destination, with huge cities, vast wildernesses, crashing waterfalls and the greatest assortment of plant and animal life on earth. It's no wonder that locals claim '*Deus e brasileiro*' (God is Brazilian).

Best time to visit
November to April on the coast, and May to September in the Amazon and the Pantanal

Top things to see
- Ipanema Beach, Rio's fabled stretch of coastline
- The enchanting colonial centre of Salvador
- The thunderous Iguaçu Falls
- A football match at Rio's Maracanã Stadium, Brazil's great temple to the national addiction
- The spectacular island of Fernando de Noronha, with gorgeous beaches and world-class diving

Top things to do
- Watch wildlife in the Pantanal, home to the greatest concentration of fauna in the New World
- Peer out over the world's most famous rainforest from a canopy tower in the Amazon
- Ride one of the continent's most scenic train routes between Curitiba and Paranaguá
- Join the mayhem of Brazil's biggest street party at Carnaval in Rio, Salvador or Olinda

Getting under the skin
Read: Oswald de Andrade's *Manifesto Antropófago* (Cannibal Manifesto), an essay on Brazilian identity
Listen: to the bossa nova grooves of Joao Gilberto; or the slick Latin R&B of Anitta
Watch: Walter Salles' poignant, Academy Award–winning *Central do Brasil* (Central Station)
Eat: *feijoada* (black bean and pork stew); or *moqueca* (Bahian fish stew with coconut milk)
Drink: *açaí* (the juice of an Amazonian berry); caipirinhas made with *cachaça* (a sugar-cane alcohol)

In a word
Tudo bem? (All's well?)

Celebrate this
Brazil gets more than 80% of its energy from renewables – most comes from hydroelectricity, with biomass and wind growing rapidly.

Random fact
Brazil's highest mountain, Pico da Neblina (2994m) was not discovered until the 1950s.

Brunei

Hemmed in by two Malaysian provinces at the north end of Borneo, adventurous travellers still drift through this tropical outpost to explore the jungles that flank the Brunei and Temburong rivers, home to proboscis monkeys, hornbills and dotted tribal villages. The contrast between the spotless modern capital, Bandar Seri Begawan (BSB), and the untamed rainforest at the city limits is just one of the country's many contradictions. Tiny Brunei is a nation built around a single personality – the enigmatic Sultan Hassanal Bolkiah. Famous for his addiction to sports cars and his hard-line interpretation of Islamic law, the sultan presides over one of the richest nations in Asia, thanks to the oil fields in the South China Sea.

Best time to visit
March to April for dry, warm days

Top things to see
- Proboscis monkeys swinging through the trees on the outskirts of the capital, Bandar Seri Begawan
- The gleaming gilded domes of the Omar Ali Saifuddien Mosque and Jame'Asr Hassanil Bolkiah Mosque
- The Brunei Darussalam Maritime Museum, rammed with treasures from a 500-year-old shipwreck
- A chariot fit for a sultan at the Royal Regalia Museum
- Exotic fruit and local delicacies at the Tamu Kianggeh food market

Top things to do
- Take a boardwalk stroll through the atmospheric stilt villages of Kampung Ayer
- Zip along rainforest rivers on the boat trip from BSB to Bandar
- Scan the jungle canopy for monkeys, snakes and hornbills in Ulu Temburong National Park
- Eat your way around the night-time hawker food markets of BSB
- Take an overnight forest trek with Iban tribespeople in Batang Duri

Getting under the skin
Read: *Some Girls: My Life in a Harem*, by Jillian Lauren, recounting her time as a guest of the sultan
Listen: to the subtly subversive hard rock of D'Hask and Eda Brig
Watch: *Yasmine*, directed by Brunei's first female director, Siti Kamaluddin, a coming-of-age tale about a young female *silat* (martial arts) champion
Eat: *ambuyat* (a thick gloop made from sago starch), often described as 'edible glue'
Drink: anything but alcohol; locals go big for juices, *kopi* (coffee) and *teh tarik* ('pulled' tea)

In a word
Panas (Hot)

Celebrate this
Despite cultural controversies, Brunei has earned plaudits for environmental conservation; some 44% of the country's virgin rainforest is protected from development, one of the highest levels of protection in the world, providing a home for thousands of rare species.

Random fact
As well as owning 7000 cars, the sultan of Brunei has his own Boeing 747 (which he often pilots himself), kitted out as a flying palace.

1. Omar Ali Saifuddien mosque in Brunei's capital is surrounded by a lagoon complete with ceremonial boat

2. Candles are part of a traditional Malay wedding

Best time to visit
May and June, September and October

Top things to see
- The hilltop Tsarevets Fortress in Veliko Târnovo
- The near-perfectly preserved Thracian tomb of Sveshtari
- The 2000-year-old Roman amphitheatre in the heart of Old Plovdiv
- The bluff-top ruins looking over the Black Sea at Kaliakra Cape
- Mountain-framed Rila Monastery, Bulgaria's most sacred site
- The 'UFO of Buzludzha', a surreal Soviet relic that looms over Stara Zagora

Top things to do
- Sip famous wines in Melnik, a tiny village with a 600-year-old wine culture
- Walk in the fragrant Valley of Roses near Kazanlâk, which bursts into flower each year in May
- Hike up to the imposing Shipka Monument, commemorating a battle against the Ottomans, for an incredible mountain panorama
- Hit the slopes in Bansko, Bulgaria's premier winter-sports town
- Search out less developed Black Sea beaches, like those south of Sinemorets

Getting under the skin
Read: *Border: A Journey to the Edge of Europe* by Kapka Kassabova, a travelogue about the region where Bulgaria, Greece and Turkey meet; or *The Physics of Sorrow* by Georgi Gospodinov, a bestseller that explores empathy
Listen: to traditional music with the *gaida* (Bulgarian bagpipe); or *chalga* disco music
Watch: *Godless*, Ralitza Petrova's story about a nurse who sells elderly patients' identities
Eat: cool *tarator* (yoghurt soup with cucumber) to beat summer temperatures; or hearty *kavarma* stews to warm up in winter
Drink: local wines that utilise local grapes such as the Mavrud or Rubin, in the birthplace of the god of wine, Dionysus

In a word
Oshte bira, molya (Another beer, please)

Celebrate this
Early Cyrillic script was created in Bulgaria, replacing Glagolitic script, and was adopted by various Slavic languages. With Bulgaria's accession to the EU, Cyrillic became the third official script of the European Union, alongside Latin and Greek.

Random fact
Bulgaria is a leading exporter of rose oil; the oil-yielding rose was most likely brought to the Valley of Roses from the Middle East.

1. Relaxing in a salt-water bath on the Black Sea

2. The now-abandoned Buzludzha Monument was once the headquarters of Bulgarian communism

3. Tsarevets was a Bulgarian stronghold until 1393

WILL SANDERS // LONELY PLANET // MATT MUNRO // LONELY PLANET // SJHAYTOV // GETTY IMAGES

B **CAPITAL** SOFIA // **POPULATION** 6.9 MILLION // **AREA** 110,879 SQ KM // **OFFICIAL LANGUAGE** BULGARIAN

Bulgaria

Despite its Roman ruins and Thracian tombs, sun-kissed vineyards and glacial lakes, Bulgaria is all too often eclipsed by famous neighbours Greece and Turkey. But in this underrated Balkan nation you'll discover mountain ranges offering misty hiking trails and cheap ski slopes, the Black Sea's golden beaches, and towns with 19th-century National Revival-era architecture guaranteed to inspire. While part of the EU, Bulgaria still half looks back to traditions like sheep yoghurt stands, carpet weaving, folk music and around 160 Bulgarian Orthodox monasteries, many of which survived 800 years living under the 'yoke' of Ottoman rule.

1. Capital city Ouagadougou has a population of about 2.5 million

2. The villages of the Gourounsi people are made from clay, straw and dung then painted by women

3. Burkina Faso is a youthful nation; the median age is 17 years

B CAPITAL OUAGADOUGOU // POPULATION 20.8 MILLION // AREA 274,200 SQ KM // OFFICIAL LANGUAGE FRENCH

Burkina Faso

Embraced in the heart of West Africa, Burkina Faso has long had a reputation for its warmth and friendliness, particularly to visitors. While those welcoming feelings have changed little among the diverse Burkinabé population, the influx of terrorism has made travel here less feasible for the time being. Meanwhile, life here goes on for those who call it home: the who's who of the Sahel's peoples still crowd into GoromGorom's Thursday market; artists and musicians continue to congregate in the vibrant cities of Ouagadougou and Bobo-Dioulasso; the Gourounsi people in Tiébéle carry on decorating their traditional cob homes in eye-pleasing geometric patterns; and elephants, antelopes and crocodiles remain patrolling the depths of Réserve de Nazinga.

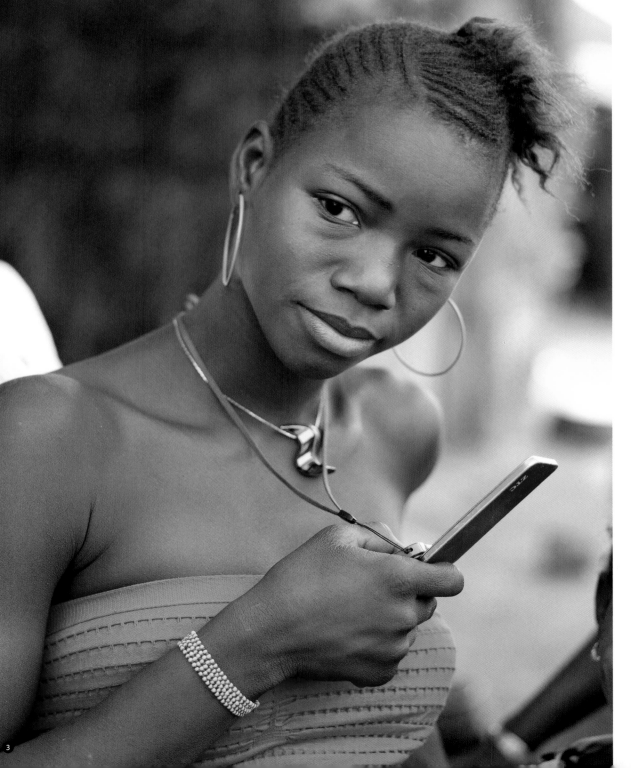

Best time to visit
November to February

Top things to see
• The tree-lined streets of Bobo-Dioulasso
• Fespaco film festival in Ouagadougou from February to March in odd-numbered years
• The colourfully painted, fortress-like houses in the heart of Gourounsi country, Tiébélé
• The waters cascading down the Karfiguéla waterfalls after the rainy season
• The 1000-year-old walls of the Ruins of Loropéni

Top things to do
• Hike through the other-worldly rock formations of the Sindou Peaks
• Track elephants and other wildlife through the Réserve de Nazinga
• Trace the Mossi Empire's history in Ouagadougou
• Explore Lobi country, an outpost of animist culture

Getting under the skin
Read: *Burkina Faso: A History of Power, Protest, and Revolution*, the story of the 2014 protests that paved the way for fair elections, by Ernest Harsch
Listen: to anything by the Volta Jazz group; the eponymous *Victor Démé* from Bobo-Dioulasso's thriving live music scene
Watch: *Burkinabè Rising*, a Iara Lee documentary showcasing the nation's nonviolent resistance
Eat: *riz sauce*, rice with sauce which could be *arachide* (groundnut) or *graine* (palm oil nuts)
Drink: Ouaga 3000 and Burkinabè La Go Fresh, two top-rated local beers

In a word
La fee bay may? (How are you?)

Celebrate this
Lassina Sawadogo, and the thousands of Burkinabé people, who said no to President Blaise Compaore in 2014 when he tried to extend his 27 years in power.

Random fact
In the Mòoré dialect, *Ouagadougou* translates to 'You are welcome here at home with us'.

Best time
June to September, when the skies are their driest

Top things to see
- The four tumultuous tumbles of the nation's greatest waterfall, Chutes de la Karera
- The ritual dance of the royal drum – powerful and synchronised, with heroic poetry and traditional songs
- The warm waters of Lake Tanganyika lapping on the soft white sands of Saga Beach
- Marked by a pyramid, the southernmost source of the Nile at the stream of Kasumo
- *Kivivuga amazina*, an improvisational contest of poetry by cattle herders

Top things to do
- Track chimpanzees through the rainforest in Parc National de la Kibira
- Spot hippos, crocs and antelopes splashing on the shoreline within Parc National de la Rusizi
- Join in the dining and frenetic nightlife of Bujumbura
- Play some football – Burundians are fanatical about it

Getting under the skin
Read: *Strength in What Remains: A Journey of Remembrance and Forgiveness*, Tracy Kidder's Pulitzer Prize–winning novel about one man's survival of the Burundi civil war; *Baho!* by Roland Rugero, translated by Christopher Schaefer
Listen: to Khadja Nin's album *Sambolera*; *Les Tambourinaires du Burundi: Live at Real World*
Watch: *Uncertain Future*, Eddie Munyaneza's documentary about the 2015 unrest and its fallout
Eat: *impeke* (a cereal made from corn, soya beans and sorghum); patisseries; fresh fish
Drink: a cold bottle of Primus beer, one of the many churned out of the national brewery

In a word
Bwa (Hello, in Kirundi)

Celebrate this
Lydia Nsekera: the first woman to be president of the Burundi Football Federation and the current head of the International Olympic Committee's Women in Sport Commission.

Random fact
Traditionally, from 8am to 11am every Saturday Burundi business ceases. The reason? *Ibikorwa rusangi* – it's a time when Burundians are required to lend a hand on community projects for the greater good of their country.

1. The Chutes de la Karera in Rutana Province are at their best in the wet season from October to January

2. Lake Tanganyika, the world's second-deepest lake, is shared by four nations: Burundi, Tanzania, Zambia and the Democratic Republic of the Congo

3. The ensemble Drummers of Burundi play at royal ceremonies

B // **CAPITAL** BUJUMBURA // **POPULATION** 11.9 MILLION // **AREA** 27,830 SQ KM // **OFFICIAL LANGUAGES** KIRUNDI, FRENCH & ENGLISH

Burundi

Although now lead by the first democratically elected president since the start of its 12-year civil war in 1994, Burundi is still a nation in conflict. There are claims that the 2020 poll, conducted during the Covid-19 pandemic, was rigged to facilitate the transfer of power from one former Hutu militia leader to another. While similar Hutu-Tutsi conflicts have been soothed in Rwanda by removing historical tribal labels, Burundi has failed to follow its alternative chosen path of open dialogue and good-hearted debate. If this struggling, often off-limits nation eventually gains stability, its jungle-clad volcanoes will offer escapades tracking chimpanzees, and its Lake Tanganyika beaches will soothe the soul. Encounters with Burundians are rewarding lessons in life – they prove that where there are lows, there still can be highs.

1. Cidade Velha, Cabo Verde's oldest settlement, is a Unesco World Heritage Site at the southern tip of Santiago

2. In the north's mountains is the village of Fontainhas

3. The port of Mindelo is famed for its Brazilian culture and music

C CAPITAL PRAIA // POPULATION 583,255 // AREA 4033 SQ KM // OFFICIAL LANGUAGE PORTUGUESE

Cabo Verde

Hewn from wind, fire and water, the African islands of Cabo Verde – found floating some 500km west of Senegal – continue to be shaped by nature's forces. And for those who visit it's hard not to be overwhelmed by the raw power of the spectacle. Whether climbing active volcanoes, exploring dune fields, trekking past craggy peaks, cycling through verdant valleys, lounging on white-sand beaches or swimming in sparkling bays, nature is always at the fore. Yet step into Cabo Verde's seaside villages and the culture is no less compelling, with a complicated blend of Portuguese and African influences in the islands' food, architecture and world-famous music. The overall effect is like nowhere else on Earth.

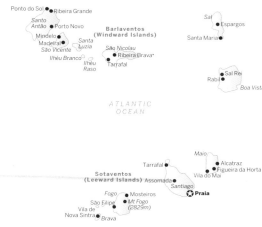

Best time to visit
October to August

Top things to see
- The Unesco-listed remnants of West Africa's first European settlement, Cidade Velha
- Mindelo, the prettiest city with a moon-shaped bay, stark mountains and a lovely old quarter
- São Filipe, a gorgeous town of historical architecture, set high on the cliffs
- The sandy beaches, desert plains and verdant, mountainous interior of Santiago

Top things to do
- Summit Pico do Fogo (2829m), Cabo Verde's highest point, for astounding vistas
- Kitesurf off the island of Sal
- Hike the precipitous cliffs and canyons of Santa Antão, the nation's most spectacular island
- Wade into the Mardi Gras festivities in Mindelo
- Enjoy the sounds of *mornas* (mournful music) in the seaside towns of São Vicente

Getting under the skin
Read: poet Jorge Barbosa's *Arquipélago*, which is laden with melancholic reflections on the sea; *The Madwoman of Serrano* by Dina Salústio
Listen: Cesária Évora's *mornas* and *coladeiras* (sentimental love songs); *Sempre Verão*, a mostly Portuguese and Creole album by Nelson Freitas
Watch: *Fintar o Destino* (Dribbling Fate) by Fernando Vendrell, a tale of a young footballer
Eat: the national dish *cachupa* (a tasty stew of beans, corn and meat or fish)
Drink: *Grogue* (sugar-cane spirit); Strela (local beer); and white or rosé wines from Fogo

In a word
Ta bon (I'm fine)

Celebrate this
Cesária Évora, known as the 'barefoot diva' for performing without shoes, passed away in 2011, but her influence on Cabo Verde culture persists.

Random fact
The long-eared bat is Cabo Verde's only native mammal.

C CAPITAL PHNOM PENH // POPULATION 16.9 MILLION // AREA 181,035 SQ KM // OFFICIAL LANGUAGE KHMER

Cambodia

In tourism terms, Cambodia was a late developer but the years since the end of its long insurgency have seen the country emerge as a key stop on the Southeast Asia backpacker trail. Much of the credit goes to the temples of Angkor, truly one of the wonders of the world. But there's more to Cambodia: stunning beaches at Sihanoukville and Koh Rong; boutique hotels and fine dining in the capital, Phnom Penh; tribal culture and wonderful wildlife in the rugged far north. Wherever you venture, there's no ignoring Cambodia's recent history, which is still framed in terms of the Vietnam War, the terrible years of the Pol Pot regime, and the slow return to peace after those dark times. Its future looks far brighter.

Best time to visit
November to January (the dry season)

Top things to see
- A vision of heaven executed in carved stone at temple-tastic Angkor
- The splash of Irrawaddy dolphins in the Mekong River at Kratie
- Rarely seen Angkor-era temples in remote Preah Vihear Province
- Mist swirling around a forgotten French hill station in Bokor National Park
- The heart-rending Khmer Rouge–era displays at Tuol Sleng and Choeung Ek in Phnom Penh

Top things to do
- Rent a motorcycle and navigate the incredible tide of motorcycle traffic in Phnom Penh
- Ride the Mekong ferry through stunning riverside scenery between Siem Reap and Battambang
- Cheer on the racing boat crews during the annual Water Festival
- Trek through steamy and uncharted jungles in Virachay National Park
- Let the surf tickle your toes on the beach at Koh Rong

Getting under the skin
Read: the harrowing memoirs *Stay Alive, My Son*, by Pin Yathay and *First They Killed My Father* by Loung Ung
Listen: to the surf rock of Ros Sereysothea; or the diverse pop and rock sounds of Cambodia's Original Music Movement
Watch: Roland Joffé's harrowing war story, *The Killing Fields*; Francis Ford Coppola's epic *Apocalypse Now*; and John Pirozzi's homage to pre-war Cambodian rock and roll, *Don't Think I've Forgotten*
Eat: *pleah* (hot and sour beef salad) or *kyteow* (rice-noodle soup) – or bite into a deep-fried tarantula
Drink: the thick and delicious fruit smoothies known as *tukalok*

In a word
Niak teuv naa? (Where are you going?) – something visitors are asked constantly!

Celebrate this
In 2020, a rat called Magawa was decorated with the animal equivalent of the George Cross for his prolific work sniffing out landmines and other pieces of unexploded ordnance left behind from the Cambodian Civil War. The rat retired in 2021.

Random fact
Pol Pot formulated his radical Marxist ideas while studying for an electronics degree in Paris.

1. The Angkor Wat temple complex is the world's largest religious site

2. Tuol Sleng Genocide Museum was a high school that was turned into a place of torture and executions

3. Koh Rong island combines tropical jungle with beaches

Best time to visit
November to February

Top things to see
- Chimpanzees fishing for termites with twigs in the ancient rainforests of Korup National Park
- Sea turtles nesting on the beautiful beach of Ebodjé
- The fascinating Sahelian town of Maroua and its colourful market
- Yaoundé, arrayed across seven hills and brimful of life
- Elephants, giant forest hogs, buffaloes, bongos and more in Lobéké National Park

Top things to do
- Shop for traditional bronze and beaded crafts at the Village des Artisans in Foumban
- Climb to the mist-shrouded summit of Mt Cameroon (4095m), West Africa's highest peak
- Depart Bamenda to explore the verdant landscapes and traditional kingdoms along the Ring Road
- Chill on a volcanic-sand beach in Limbe, with the slopes of massive Mt Cameroon rising behind you
- Go truly wild in the remote jungles of the Dja Faunal Reserve

Getting under the skin
Read: *Your Madness, Not Mine: Stories of Cameroon* by Juliana Makuchi Nfah-Abbenyi; *Limbe to Lagos: Nonfiction from Cameroon and Nigeria*, personal essays about life from ten young writers
Listen: to any Afrobeats by Magasco (aka Bamenda Boy); Dibango's *Soul Makossa*, one of Africa's most influential albums
Watch: *Afrique, Je Te Plumerai* (Africa, I Will Fleece You) by Jean-Marie Teno, an outstanding documentary about modern Cameroon; *A Man for the Weekend*, a Cameroonian rom-com
Eat: *fufu* (mashed yam, corn or plantain) with *ndole* (a bitter-leaf-and-smoked-fish sauce); or *suya* (beef cooked on outdoor grills)
Drink: *Bil-bil* (beer made from millet, corn or sorghum); local coffee; and milky-white palm wine

In a word
No ngoolu daa (Hello, in Fulfulde)

Celebrate this
Andre Blaise Essama: an activist who has been on a mission to purge his country of colonial-era symbols since 2003; he wants to replace them with heroes from Cameroon and elsewhere in Africa.

Random fact
Lake Nyos, one of two 'exploding lakes' in Cameroon, is considered the most deadly on earth – it killed around 1700 people in 1986 before efforts were made reduce its volcanic gases.

1. Yaoundé became the capital city of what is now Cameroon in 1922

2. Korup National Park in the southwest contains remnants of the great Atlantic Coastal Forest

3. The Chutes de la Lobé near this beach empty straight into the sea

C **CAPITAL** YAOUNDÉ // **POPULATION** 27.7 MILLION // **AREA** 475,440 SQ KM // **OFFICIAL LANGUAGES** FRENCH & ENGLISH

Cameroon

At the crossroads of West and Central Africa, Cameroon's tendril-like borders encompass a spectrum of African landscapes: tropical beaches in the south, steamy rainforests of the interior and the semi-deserts of the Sahelian north. Oh, and then there is the string of active volcanoes cutting a swathe inland from the coast. Its human geography is similarly dramatic and diverse: from 263 peaceful ethnic groups and a wealth of traditional kingdoms still holding sway, to cultural remnants of German, French and British colonisers. Cameroon is truly a humid mosaic like no other. Sadly, incursions of Boko Haram and Islamic State West Africa violence have also been added to the mix of late, making parts of the country off limits to travellers.

Best time to visit

March to November, except in the north where winter comes early (October) and leaves late (April)

Top things to see

- Black bears, grizzly bears, moose and more in the wilds across the country
- Québec City's Old Town, a Unesco World Heritage Site exuding historical romance
- The Haida Gwaii archipelago, where the rich Haida culture is thriving again
- The world-class restaurants and chic boutiques of downtown Toronto
- Newfoundland's Northern Peninsula, which blends icebergs, cliffs and the odd Viking artefact

Top things to do

- Hike the craggy peaks and alpine meadows of Banff National Park
- Carve up some fresh powder at one of the Canadian Rockies' renowned ski resorts
- Sail the whale- and dolphin-filled Inside Passage along British Columbia
- Take a coast-to-coast train trip across the whole country
- Tuck into dim sum in Vancouver, where one in five residents have Chinese heritage

Getting under the skin

Read: *Selected Stories* by Nobel Prize–winner Alice Munro, largely set in rural Ontario; or Ann-Marie MacDonald's Nova Scotia epic, *Fall on Your Knees*
Listen: to Leonard Cohen, Neil Young, Drake and Arcade Fire
Watch: *Atanarjuat: The Fast Runner*, an Inuit legend told in the Inuktitut language; or *Les Démons*, a scary, funny look at childhood in Montréal
Eat: fresh seafood; maple syrup; poutine (fries with gravy and cheese curds)
Drink: many superb wines from the Okanagan Valley in southern British Columbia

In a word

Eh? (bilingual and all-purpose, eg 'Nice day, eh?')

Celebrate this

Canadians love (ice) hockey, a sport that has crossed gender, language and race barriers since it developed in the 19th century from stick and ball games brought by the British and the First Nations Mi'kmaq.

Random fact

Every year the British Columbian town of Nanaimo holds a bathtub race, where competitors speed across the harbour in boats formed from bathtubs.

1. A humpback whale breaches near Vancouver Island; whale populations have recovered strongly and there's a good chance of seeing them on the west or east coasts

2. Le Château Frontenac and Old Québec

3. Moraine Lake in Banff gets its colour from light refracting off fine rock particles

DANIEL BEARHAM | GETTY IMAGES // AURÉLIEN POTTIER | GETTY IMAGES // JUSTIN FOULKES | LONELY PLANET

C **CAPITAL** OTTAWA // **POPULATION** 37.7 MILLION // **AREA** 9,984,670 SQ KM // **OFFICIAL LANGUAGES** ENGLISH & FRENCH

Canada

You can literally lose yourself almost anywhere in Canada. The world's second-largest country has more gorgeous and remote corners than you could ever count or visit. From the glaciers of Kluane National Park in the Yukon to Nova Scotia's Cape Breton Highlands, the natural wonders never cease. When you're ready for civilisation you'll find cities that are among the world's most genteel and enjoyable – and multicultural, combining elements of Britain, France, Asia and almost every other world region. Its indigenous peoples have robust cultures that enrich the nation too, with art that honours both the glories of nature and aeons-old traditions. All in all, Canada offers an intoxicating mix of people and landscapes.

C CAPITAL BANGUI // POPULATION 6.0 MILLION // AREA 622,984 SQ KM // OFFICIAL LANGUAGE FRENCH

Central African Republic

Beneath its soil sit diamonds, gold, oil and uranium, yet the Central African Republic (CAR) has one of the planet's poorest populations. The reason? Centuries of exploitation. First, its sophisticated society (thousands of years in the making) was shattered by the trade in humans. Then the desperate remnants were catapulted into the harsh yoke of French colonial rule, only to be subjected to agonisingly egocentric governments after independence. Relative stability existed between 2008 and 2012 before the country was thrust back into civil war. Since then neither a UN-led peacekeeping mission nor a new referendum-backed constitution have eased tensions, making the CAR too dangerous to visit. Beyond the conflict, wildlife lives in its jungles, and warmth, generosity and pride steadfastly survive in the hearts of its people.

Best time
November to April, its dry season

Places of interest
- Chutes de Boali, a 50m-high waterfall that bursts to life in the rainy season
- The ruined palace of former 'Emperor' Bokassa at Berengo
- Megalithic stone monuments that dot the landscape around the market town of Bouar
- Bangui's scenic river frontage and its old town
- Dzanga-Sangha Reserve and its thousands of western lowland gorillas
- The Congo Basin, where rare forest elephants roam beneath the tree canopy

Local customs
- Baka (a Pygmy tribe) summon forest spirits as they hunt *mboloko* (blue duiker)
- Traditional storytelling, with the folklore being sung to the music of drums, *ngombi* (harp) and *sanza* (guitar)
- Drinking palm wine and dancing the *gbadoumba* and *lououdou*
- Initiation rites and traditional religious practices

Getting under the skin
Read: *Making Sense of the Central African Republic*, a collaboration of experts discussing the nation's recent history of uprising, insecurity and international and regional intervention
Listen: to anything by Canon Star, a 10-strong band that is known for its live performances
Watch: *Camille*, a story about a 26-year-old French reporter who was killed in the CAR
Eat: plenty of manioc – *ngunza* (manioc leaf salad) and *gozo* (manioc paste) are local favourites
Drink: locally brewed banana or palm wine

In a word
Bara ala kwe (Hello, in Sango)

Celebrate this
Henriette and Jean-Philippe Idjara, and the hundreds of other Central African foster parents caring for children orphaned during the nation's recent violence.

Random fact
Jean-Bédel Bokassa wasn't happy with just being president of the country, so in 1976 he converted the Central African Republic into the Central African Empire; most of the bill for the US$20 million coronation to make him emperor was footed by France.

1. A young forest elephant in Dzanga Bai, a reserve in the Unesco World Heritage Site Sangha Trinational

2. The hunter-gatherer Baka people live in the forests of the southwest

3. The M'bari River tumbles over the Chutes de Boali

Best time
December to mid-February

Top things to see
- Increasing populations of elephant, buffalo and roan antelope in Zakouma National Park
- The Sahara and its powdery sands, complete with elegant, perfectly preserved shells of aquatic molluscs
- The beautifully painted mud-brick houses of the Sao people in the village of Gaoui
- The cooling waters of the dramatic Guelta d'Archei, which hundreds of camels (and a few crocodiles) share
- Hippos lurking in the Chari River in the nation's capital, N'Djaména

Top things to do
- Trek to the peak of Emi Koussi's cratered summit, which looms over the Sahara
- Take an expedition into the dramatic desert and mountain scenery of the Ennedi region
- Explore the green and pleasant banks of the riverside town of Sarh
- Head out onto the waters of Lake Chad from the village of Bol

Getting under the skin
Read: *Living by the Gun in Chad: Governing Africa's Inter-Wars: Combatants, Impunity and State*, by Marielle Debos
Listen: to Tibesti, a group who popularised the rhythmic *sai* sound
Watch: *A Screaming Man*, a present-day tale of a former swimming champion; *Abouna*, a gripping story following two boys' misguided search for their father
Eat: *nachif* (finely minced meat in sauce); *salanga* and *banda* (sun-dried/smoked fish)
Drink: a bottle of Gala beer, from the brewery in Moundou

In a word
Lale (A warm greeting in southern Chad)

Celebrate this
The rejuvenation of wildlife within Zakouma National Park: the populations of elephant, buffalo, roan antelope, Lelwel's hartebeest and kordofan giraffe are all on the rise.

Random fact
Chad may eventually be without its namesake, Lake Chad – desertification and human water use over the past 50 years have reduced its size from over 26,000 sq km to less than 1350.

1. Zakouma National Park is home to the rare subspecies the Kordofan giraffe

2. The roof of the Notre Dame de la Paix cathedral in N'Djaména was destroyed by war in 1980

3. Wadi Archei is an important source of water for camels

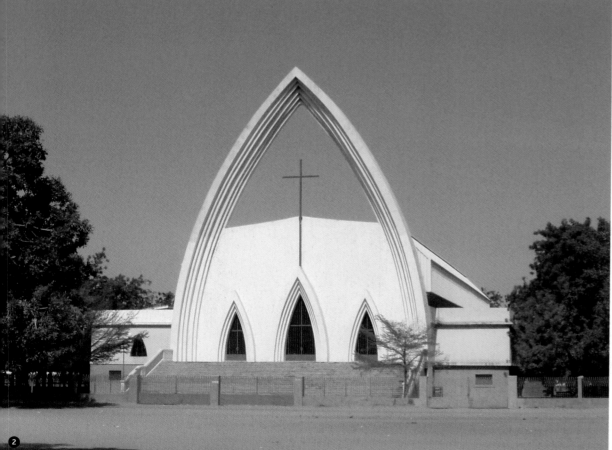

C CAPITAL N'DJAMÉNA // POPULATION 16.9 MILLION // AREA 1,284,000 SQ KM // OFFICIAL LANGUAGES ARABIC & FRENCH

Chad

With foundations built on ethnic and religious conflict, Chad continues to teeter on the brink. Despite oil production and reserves of gold and uranium, poverty is the only true unifier between the Arab-Muslim north and the Christian and animist south. There have been some positive moves of late, such as the country working cooperatively with regional neighbours and France to halt the expansion of Islamic militant groups (including Boko Haram). Yet, terrorism from these groups is still a huge issue in areas, as is the southern march of the Sahara, which continues to desertify more and more of the country. Meanwhile, beyond the violence, growing political protests and shifting sands, life abounds. Chad, like its people, contains glimpses of surreal beauty with harsh doses of reality.

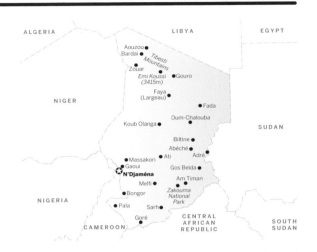

1. Charles Darwin passed through the foggy port of Valparaíso in 1834 on his way to the Galápagos Islands, learning first-hand about earthquakes in the process

2. Parque Nacional Torres del Paine is perhaps the greatest protected area in South America

3. There are almost 1000 stone moai on Easter Island, one of the world's most remote inhabited islands

C CAPITAL SANTIAGO // POPULATION 18.2 MILLION // AREA 756,102 SQ KM // OFFICIAL LANGUAGE SPANISH

Chile & Easter Island

Long and slim, Chile spreads down a coastal strip of South America between the Pacific and the Andes, spanning half the length of the continent from the earth's driest desert to the largest glacial fields outside the poles. The arid plains, mountains, port cities and resort towns of the north give way to lush valleys and vineyards in the centre of the country, as well as the capital, Santiago, its skyline ringed by peaks and pierced by increasingly ambitious building projects.Then, in increasing splendour, come Andean summits, forests, lakes and the mist-cloaked, mystery-soaked islands of the lonely, lacerated southern coast – not to mention the mysteries of Easter Island. No surprises, with such scenery, that the country has inspired a glut of renowned writers, such as Roberto Bolaño and Isabel Allende, as well as a regular influx of adventure-seekers.

Best time to visit
Year-round in the north, November to April in the south, June to August for skiing

Top things to see
- The lakes, volcanoes and ancient *alerce* trees of the Lake District
- Santiago's glittering skyline from the powder-clad ski slopes above it
- The thousands of islands studding the end-of-the-earth Strait of Magellan
- The desert around San Pedro, home to lunar valleys, geysers and salt lakes
- Rano Raraku, the mountain quarry of the mammoth Easter Island statues

Top things to do
- Sip wines in the Central Valley's vineyards
- Trek under the sharp spires of Torres del Paine
- Feast on fresh oysters in the archipelago of Chiloé
- Mosey with pack horses through the river valleys of northern Patagonia

Getting under the skin
Read: Gabriel García Márquez's exposé on Chile under Pinochet's dictatorship, *Clandestine in Chile*
Listen: to groundbreaking '90s rock outfit La Ley
Watch: Sebastián Lelio's *Gloria*, about a middle-aged divorcee's adventures on the dating scene
Eat: conger eel soup; crab casserole; a *completo*: hotdog with a hotchpotch of sauces and seasonings
Drink: a tart pisco sour

In a word
Bacán! (Cool!)

Celebrate this
Chile's claim to be a 'country of poets' has merit: Gabriela Mistral and Pablo Neruda both won the Nobel Prize for Literature, while Roberto Bolaño is one of South America's greatest authors.

Random fact
Thanks to its clear skies and low light pollution, the Atacama Desert has Earth's best stargazing; its ALMA observatory is investigating the birth of stars.

Best time to visit
March to May and September to November

Top things to see
- The Great Wall, actually a series of walls, forming an imperious line across central China
- Běijīng's Forbidden City, China's grandest imperial palace
- The sculpted faces of Xī'ān's Army of Terracotta Warriors
- Snow-capped peaks rising above the 2km-deep cliffs of Tiger Leaping Gorge
- The swirling landscape of the Lóngjǐ Rice Terraces

Top things to do
- Take an overnight train journey from province to province to grasp China's enormous scale
- Ride a bamboo raft between jagged karst mountains near Yángshuò
- Sample China's culinary landscape, from fiery Chóngqìng hotpot to crispy Peking roast duck
- Stroll past the kite flyers and British-era office buildings on Shànghǎi's Bund
- Wander into the desert where the Great Wall collapses to dust near Jiayuguan and Dunhuang

Getting under the skin
Read: *Country Driving: A Chinese Road Trip*, Peter Hessler's engaging travel memoir; and *Tiger Head, Snake Tails* by Jonathan Tenby
Listen: to the platinum-selling pop songs of Jacky Cheung, Aaron Kwok, Andy Lau and Leon Lai – the 'Four Heavenly Kings' of Cantopop
Watch: Zhang Zimou's *Raise the Red Lantern*, Fei Mu's *Spring in a Small Town*, Stephen Chow's *Kung-Fu Hustle*, and Frant Gwo's *The Wandering Earth* to appreciate the remarkable range of Chinese cinema
Eat: Sìchuān dishes flavoured with 'flower pepper', a fiery spice unrelated to chillies or black pepper
Drink: *chá* (tea) at a traditional teahouse; or *baijiu*, a whole family of rice- and grain-based liquors

In a word
Chīfàn le ma? (Have you eaten?)

Celebrate this
The mantle of legendary human rights campaigner Liu Xiaobo (1955-2017) has been taken up by a new generation of activists such as teenaged climate activist Howey Ou and Uighur campaigner Ilham Tohti, pushing forward the movement for political and minority rights.

Random fact
Among other things, the Chinese invented paper, printing, gunpowder, the compass and the umbrella.

1. The Hall of Supreme Harmony in Běijīng's Forbidden City

2. The city of Hong Kong viewed from Victoria Peak

3. Practising the Chinese martial art tai chi

4. The sandstone pillars of Zhangjiajie National Park in the Wulingyuan Scenic and Historic Interest Area

C CAPITAL BĚIJĪNG // **POPULATION** 1.4 BILLION // **AREA** 9,596,961 SQ KM // **OFFICIAL LANGUAGE** MANDARIN

China

A magnificent sprawl of cities, bamboo forests, deserts, rice terraces and mountains, the world's most populous nation is a place of history, culture and creativity, but visiting poses some challenging questions: progress has come against a backdrop of human rights abuses. China is the world's largest producer and user of solar panels, but, by being the world's factory, is also – by some distance – the world's largest emitter of CO_2. Nevertheless, despite the headlong rush for modernity, China's five millennia of history is tangible everywhere. For every newly built skyscraper, there's a centuries-old pagoda, and for every fast-food franchise, there is a traditional teahouse serving hand-pulled noodles.

Best time to visit
January to March (the dry season)

Top things to see
- The cobbled lanes of Cartagena, Colombia's most romantic colonial city
- The Zona Cafetera, a coffee-growing region set against a backdrop of volcanoes
- The rebirth of Bogotá, a vibrant and style-conscious city with a burgeoning arts scene
- Laid-back Capurgana and Sapzurro, two blissfully old-fashioned settlements ringed by rainforest on the Caribbean coast
- The subterranean salt cathedral in Zipaquirá

Top things to do
- Head to the adventure capital of San Gil, with incredible rafting, caving, horseback riding and mountain biking
- Journey to Colombia's Amazonian wilderness at the Reserva Natural Zacambú
- Trek through rainforest and mountains to the ruins of the Ciudad Perdida
- Explore the archaeological sites in the rolling hills around San Agustín
- Party till dawn in Medellín, Colombia's renaissance city

Getting under the skin
Read: *Love in the Time of Cholera*, the fantastical love story by Nobel Prize–winning Colombian author Gabriel García Márquez
Listen: to pop superstar Shakira, reggaeton singer J Balvin or the Afro-Caribbean beats of Totó La Momposina
Watch: Ciro Guerra's *Embrace of the Serpent*, a mystical but passionately political exploration of the Amazon and its peoples
Eat: satisfying *arepas* (corn cakes served with cheese, pork and other toppings); *sancocho* (a soup made with meat, yucca and other vegetables)
Drink: arguably the world's best coffee: order a *tinto* (black, espresso size), *pintado* (small milk coffee) or *cafe con leche* (standard size, with more milk than coffee)

In a word
Que hubo? (What's up?)

Celebrate this
Colombia is famous for its biodiversity: the only country in South America to have Pacific and Atlantic coasts, its landscapes house around 20% of the world's plant and bird species and 10% of the world's mammals, plus more types of butterfly and orchid than anywhere else on the planet.

Random fact
Avianca, Colombia's flagship airline, was the first commercial airline founded in the Americas.

1. The Ciudad Perdida (Lost City) can be reached with a steamy hike through the Sierra Nevada de Santa Marta

2. Colonial buildings in beautiful Cartagena

3. El Cabo San Juan del Guia beach in Parque Nacional Natural Tayrona

C **CAPITAL** BOGOTÁ // **POPULATION** 49.1 MILLION // **AREA** 1,138,910 SQ KM // **OFFICIAL LANGUAGE** SPANISH

Colombia

Once-troubled Colombia is one of South America's most remarkable success stories. A decades-long civil war has been largely relegated to the past, and peace and economic growth are helping make the country a major tourist draw. Colombia's trump card is its diverse geography, which includes Andean peaks, rainforests and grasslands, supporting an astounding 10% of the world's plants and animal species. A flurry of urban renewal projects has breathed new life into Colombia's cities, and its natural wonders are now more accessible than ever. Whether you seek jungle adventures or prefer to kick back in a colonial village with a cup of Colombian coffee, you can expect the famed hospitality of the nation's people – who mix indigenous, African and European ancestry – at every turn.

CAPITAL MORONI // **POPULATION** 846,281 // **AREA** 2235 SQ KM // **OFFICIAL LANGUAGES** ARABIC, FRENCH & SHIKOMORO

Comoro Islands

Born of fire, these Indian Ocean islands have seldom come off the boil. Discontent, intrigue and ambition, much like the omnipresent lava lurking beneath this archipelago, have all erupted habitually, resulting in civilian riots and more than 20 coups since 1975. However, for visitors, the melange of Polynesian, Swahili and Arabic cultures is as intoxicating as the fragrant fields of ylang-ylang, jasmine, cassis and orange flowers. Also rooted in the islands' fertile volcanic soils are virgin rainforests, which host everything from giant bats to rare lemurs. Surrounding it all is a coast blessed with turquoise waters, colourful-sand beaches and ports laden with historical architecture.

Best time to visit
May to October (the dry season)

Top things to see
- The highest concentration of coral marine species within a small area in the world
- Fringes of faultless sand on the beaches lining the islets of Parc Marin de Mohéli
- An entire nation and its archipelago of islands spread out below you from the summit of Mt Ntingui
- Green turtles laying their eggs peacefully on the beach near Itsamia village on Mohéli
- Black-sand beaches fringed with palms on the island of Anjouan

Top things to do
- Trek up to the peaceful crater lakes of Dzialaoutsounga and Dzialandzé on Mt Ntingui
- Take on a local in the ancient African game of *bao*
- Raise a sail and ride the southeastern trade winds atop the Comoros' tropical waters
- Meander the crooked alleys of Moroni's ancient medina, admiring the intricately carved Swahili doors as you go
- Inhale the sweet fragrance wafting from one of the many the ylang-ylang distilleries

Getting under the skin
Read: *Islands in a Cosmopolitan Sea: A History of the Comoros*, by Iain Walker; *The Comoros Islands: Struggle Against Dependency in the Indian Ocean*, Malyn Newitt's outline of the region's turbulent, coup-riddled modern times
Listen: to Nawal, who mixes her island's rhythms with various musical aspects from the Indian Ocean region: sounds from India, Arabia and Persia; polyphony textures from African bantu tribes; and *dhikr* (Sufi chanting)
Watch: the films of Hachimiya Ahamada about Comoran diaspora: *Feu Leur Rêve* (The Dream of Fire), *La Résidence Ylang-Ylang* (The Ylang-Ylang Residence) and *Ivresse d'une Oasis* (Ashes of Dreams)
Eat: reasonably priced lobster – *langouste à la vanille* is particularly divine
Drink: tea spiced with lemongrass and ginger

In a word
Salama (Hello, in Comoran)

Celebrate this
Hadjira Oumouri: a former midwife and the nation's second-ever female MP, she has had success introducing bills that have advanced the rights of women in Comoros.

Random fact
A wedding, or *grande marriage* as it's known in the Comoros, can last up to nine days – the groom is expected to fund the *toirab* (celebration) that caters for the entire village.

1. Looking out towards the city of Domoni on the west coast of Anjouan island

2. Green sea turtles inhabit the ocean around the Comoros

3. A wedding ceremony takes place in the city of Moutsamudu on the island of Anjouan

①

Best time
December to February (north of equator), May to October (south of equator)

Top things to see
- The Congolese Society of Ambience-makers and Elegant Persons (aka the *sapeurs*) in their finest
- Wagenia Falls, where working fisherfolk balance deftly above the torrents on wooden platforms
- Bonobos, rare peace-loving relatives of chimpanzees, at the Lola Ya Bonobo Sanctuary
- A flood in a 347m freefall from the top of Lofoi Falls, Parc National de Kundelungu
- Automated robots – complete with waving arms and tin voices – which control traffic in the chaotic streets of Kinshasa

Top things to do
- Climb Mt Nyiragongo to look down into its crater as it spews lava skyward
- Experience the colours and chaos of Kinshasa
- Venture into Africa's oldest park, Parc National des Virunga, to encounter mountain gorillas
- Journey by barge down the Congo River as it meanders through the wilderness and past villages
- Shop for embroidered textiles, ornate cosmetics boxes and masks made by descendants of the Kuba Kingdom

Getting under the skin
Read: *King Leopold's Ghost* by Adam Hochschild; and *In the Footsteps of Mr Kurtz*, Michela Wrong's compelling look into President Mobutu's regime
Listen: to the kings of *soukous* (African rumba), Franco Luambo and Papa Wemba, Africa's equivalent of James Brown and Elvis Presley
Watch: *Virunga*, a documentary focussing on four rangers trying to save gorillas and Parc National des Virunga; *This is Congo*, Daniel McCabe's award-winning documentary on the nation's recent conflicts
Eat: *liboke* (fish stewed in manioc leaves) with the omnipresent *fufu* (manioc porridge)
Drink: Legend Extra Stout beer, or Turbo King if you prefer dark lager

In a word
Sángo níni? (How are you? in Lingala)

Celebrate this
Immaculée Birhaheka, a human rights activist who co-founded the women's rights organisation Promotion et Appui aux Initiatives Feminines (PAIF): she continues to work to combat sexual violence in the DRC and to increase women's roles in society.

Random fact
On the Congo River's southern bank, Kinshasa sits across from Brazzaville in the Republic of Congo; it's the only place in the world where two national capitals are situated on the opposite banks of a river within sight of each other.

1. Residents of the village of Wagenia have used this technique to fish the Congo River for centuries

2. In the east of the country, around 300 mountain gorillas live in the Parc National des Virunga, where the population is protected by dedicated rangers

3. Mt Nyiragongo, an active volcano in the Virunga mountains

②

CAPITAL KINSHASA // POPULATION 102 MILLION // AREA 2,344,858 SQ KM // OFFICIAL LANGUAGE FRENCH

Congo, Democratic Republic of the

When it comes to African treasures, the Democratic Republic of the Congo (DRC, formerly Zaire) has plenty – rainforests, volcanoes, rivers, mountain gorillas, mineral wealth and cultural riches. Yet, it also has more than its share of problems. Generations of its people have had to witness successive kleptocrats running government coffers dry, and – worse still – endure violent conflict, including Africa's first 'world war', which left over five million dead by its end in 2008. Yet in the past decade, this massive nation (which dwarfs Western Europe) has seen a trickle of intrepid travellers experience a range of encounters and experiences: compelling conversations with locals in cities and remote villages; interactions with silverbacks and elephants in impenetrable forests; and exhilarating hikes to overlook lava lakes. When safe and accessible, the DRC is an adventurer's dream.

1. Using a pole to pilot a pirogue is a great way to navigate the Lekoli River

2. The wetlands of Parc National d'Odzala support a huge range of animals including 444 species of bird and large numbers of western lowland gorillas

3. A forest elephant in Parc National d'Odzala; they tend to be smaller than their savannah-dwelling relatives

CAPITAL BRAZZAVILLE // **POPULATION** 5.3 MILLION // **AREA** 342,000 SQ KM // **OFFICIAL LANGUAGE** FRENCH

Congo, Republic of the

A place of legendary peoples, wildlife and jungles, the Congo's name alone stirs the adventurous spirit in every intrepid traveller. Travel north and the sultry air beneath the thick canopy of the Congo's dense jungles reverberates with the chest thumping of lowland gorillas, the pan hoots of chimpanzees, the sound of forest elephants and the hunting calls of the Mbenga people (formerly known as Pygmies). Although much more predictable than neighbouring Democratic Republic of the Congo, this nation has seen also some unrest, most recently following the turbulent 2016 elections. A 2017 ceasefire agreement has moved the country forward, as have the use of oil and timber revenues to expand infrastructure and – happily for wildlife and tourism – to rehabilitate national parks. Travel here will always be an intense experience, but the rewards are equally unforgettable.

Best time to visit
December to February (north of equator), May to September (south of equator)

Top things to see
- Chimpanzees within the coastal rainforests of Parc National Conkouati-Douli
- Herds of forest elephants congregating in Wali Bai, a clearing in Parc National Nouabalé-Ndoki
- Congolese navigating the shifting currents of the Congo River in their pirogues
- Rusty, tendril-like ridges of rock snaking through the rainforest in Diosso Gorge

Top things to do
- Walk beneath the tree canopy of Parc National d'Odzala in search of gorillas and elephants
- Explore the protected marine areas of Parc National Conkouati-Douli, home to sea turtles and rare West African manatees
- Sip a cappuccino on a terrace in Brazzaville and watch Africans, Arabs, Europeans and Asians going about their day
- Laze on a beach in Pointe-Noire

Getting under the skin
Read: *Lumières de Pointe-Noire*, Alain Mabanckou's story of his return home; *Congo Journey*, Redmond O'Hanlon's captivating travelogue
Listen: to Roga Roga, a master of Congolese *ndombolo*, also known for his *soukous* and Afro-pop
Watch: *Congo*, Frank Marshall's adaptation of the novel by Michael Crichton
Eat: fresh fish with fried bananas
Drink: the ubiquitous palm wine

In a word
Losáko (Hello, in Lingala)

Celebrate this
Elende Gilbert: whose projects to foster sustainable livings for locals in the region have included planting 40,000 cocoa seedlings.

Random fact
The Congo is the world's deepest river (up to 220m).

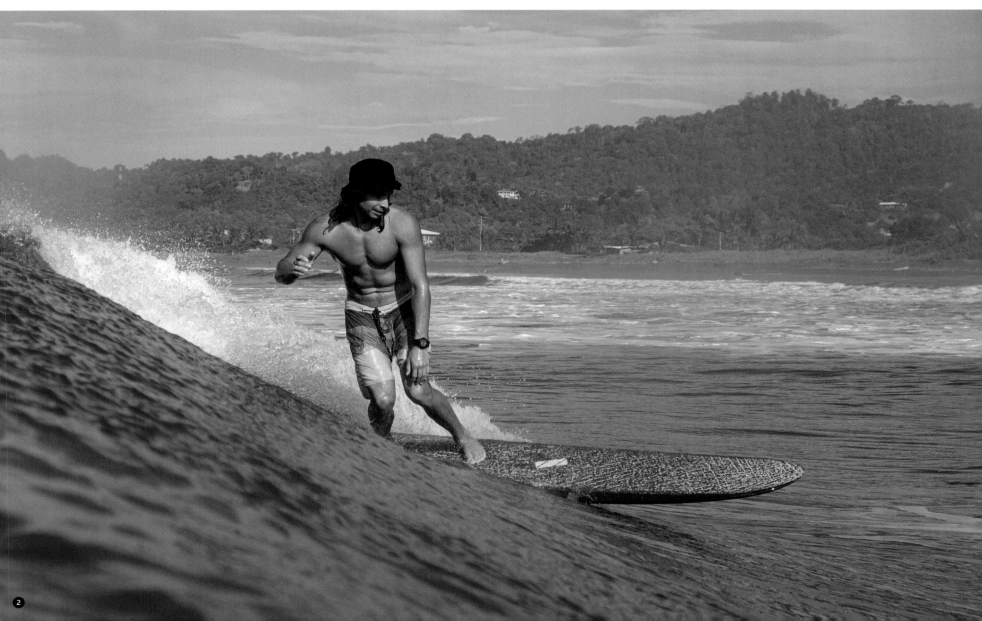

C · CAPITAL SAN JOSÉ // POPULATION 5.1 MILLION // AREA 51,100 SQ KM // OFFICIAL LANGUAGE SPANISH

Costa Rica

Costa Rica is a spot of calm in a region of strong political currents, with high education levels, stable politics and no standing army. Agriculture and finance are key industries, and tourism is a major player, with surfers and wildlife watchers coming here in droves. The rainforests are home to toucans, caiman, sloths, capuchin monkeys and butterflies. Plus, the waves are prime, the beauty is staggering and the slow pace of life is seductive. Of course playing paradise to the world has its consequences. Now two-thirds of the coast is foreign-owned and development often trumps environmental concerns. Thankfully, there's a real focus on sustainability, with the country targeting carbon-neutrality, and government incentives encouraging reforestation and conservation.

Best time to visit
December to April (the dry season); for a quieter but wetter trip, consider May to November

Top things to see
- The fantastic biodiversity of Parque Nacional Corcovado
- The windswept wilderness beaches at the tip of the Nicoya Peninsula
- The forested hills, gorgeous beaches and bold monkeys of Parque Nacional Manuel Antonio
- Leatherback sea turtles making their epic migration to Playa Grande
- Masked devils schooling the colonial Spanish in the Fiesta de los Diablitos

Top things to do
- Whiz across the cloud forest canopy on a zip line in Monteverde
- Soak in the hot springs in the shadow of the smoking Arenal Volcano
- Ride the waves at Witch's Rock
- Dance to calypso beats in Cahuita, on the steamy Caribbean coast
- Paddle a maze of jungle canals thick with wildlife in Parque Nacional Tortuguero

Getting under the skin
Read: *Costa Rica: A Traveler's Literary Companion*, edited by Barbara Ras, 26 short stories offering different perspectives on the country
Listen: to *Costa Rica: Calypso*, fun Caribbean sound from rootsy trad to pop
Watch: Miguel Gómez's *Maikol Yordan Traveling Lost*, a tale of a farmer's travels, packed with Tican humour, that became the most watched film in Costa Rican history
Eat: *casado* (a plate of meat, beans, rice and fried plantain); sea bass ceviche
Drink: palm wine, the preferred firewater of rural farmers; coffee at any local lunch counter

In a word
Pura vida (Pure life) – for thumbs up or a salutation

Celebrate this
There are concerns over tourist development and ageing infrastructure, but Costa Rica's green credentials remain world-beating and the country regularly tops global surveys. Over 99% of its electricity comes from sustainable sources, while laws protect mature forests.

Random fact
Since 1948, Costa Rica has had no armed forces.

1. Arenal stratovolcano, although one of the world's more active volcanos, has been dormant since 2010
2. There are some fine waves around Jaco on the central Pacific coast
3. Although the green-and-black poison dart frog found in Costa Rica's rainforests grows to just 5cm in length, it's highly toxic
4. Parque Nacional Corcovado in southwest Costa Rica is one of the world's most biodiverse wild places, accessed only on foot

Best time to visit
November to March

Top things to see
- Contemporary African art, skyscrapers and mouth-watering plates of Ivoirian dishes in Abidjan
- Zaouli, a traditional masked dance by the Guro people of central Côte d'Ivoire
- Yamoussoukro, the village-turned-capital with the tallest basilica in Christendom
- The stilt dancers in the village of Silacoro

Top things to do
- Soaking up the beach vibes in the coastal town of Grand Bassam
- Surf the Atlantic breakers at Assinie, or watch those doing so while lazing in a pirogue
- Hike the stunning mountains and green valleys in the region of Man
- Search for nut-cracking chimpanzees in the dense rainforests of Parc National de Taï
- Haggle for the famous Korhogo cloth in the rust-red city of the same name in northern Côte d'Ivoire

Getting under the skin
Read: *Far from my Father*, the story of a woman returning to Côte d'Ivoire following her father's death
Listen: to Alpha Blondy, the king of West African reggae; or for more recent beats, Ariel Sheney
Watch: *Night of Kings*, a story of a young thief who survives in an Abidjan prison by telling stories to other inmates; Henri Duparc's acclaimed *Bal Poussière* (Dancing in the Dust), which tackles traditional polygamy
Eat: *poisson braisé* (grilled fish) with *attiéké* (grated cassava); or *kedjenou* (chicken or guinea-fowl simmered with vegetables in a mild sauce)
Drink: Doppel Munich or Solibra Chill beers; or *bandji* (palm wine)

In a word
I-ni-cheh, I-kah-kéné (Hello, how are you? in Dioula, the market language)

Celebrate this
O'Plérou Grebet: a young Ivorian artist who created 365 emojis representing elements of West African culture; he was subsequently chosen to create the UNHCR's 2020 World Refugee Day emoji.

Random fact
Despite its name translating as 'Ivory Coast' (and its national football team being called Les Éléphants), fewer than 300 of the mighty pachyderms are estimated to live in the country.

1. A shopkeeper in French-colonial resort town of Grand Bassam on the coast

2. The Basilica of Our Lady of Peace of Yamoussoukro is thought to be the world's largest church

3. The world-class beach at Plage d'Assinie faces the Atlantic

C **CAPITAL** YAMOUSSOUKRO // **POPULATION** 27.5 MILLION // **AREA** 322,463 SQ KM // **OFFICIAL LANGUAGE** FRENCH

Côte d'Ivoire

From the confident and cosmopolitan coastal city of Abidjan to the deeply traditional regions of the rural interior, Côte d'Ivoire is 21st-century Africa in microcosm. Following almost a decade of unrest, a return to stability led the economy to blossom in 2012. Since then it has continued to boom, though there are many people who feel that they have been left out of its successes, which has caused some political instabilities, including a brief military mutiny in 2017. To the traveller, the nation is as beautiful and welcoming as always, with beaches, rainforests and mountains ready for exploration and enjoyment.

C CAPITAL ZAGREB // POPULATION 4.2 MILLION // AREA 56,594 SQ KM // OFFICIAL LANGUAGE CROATIAN

Croatia

Croatia's dazzling stretch of Adriatic bays, beaches and rocky islets is well and truly buzzing with visitors, and with good reason. In many places the Mediterranean lifestyle of harbourside strolls, seafood and liberally poured wine continues. But Croatia's post-independence energy means you'll find little time to siesta in between thought-provoking galleries and some of Europe's best summer music festivals. Croatian culture has manifold influences – Venetian merchants, Slavic folklore, rugged geography, devout Catholicism and maritime adventurism – all of which contribute to this sun-drenched land of warm stone architecture where oleanders seem to bloom year-round and the water is implausibly clear.

Best time to visit
April to June and September to October

Top things to see
- Dubrovnik, the 'pearl of the Adriatic', a limestone-walled city on the sea
- The sparkling lakes, waterfalls and walkways at Plitvice Lakes National Park
- Diocletian's Palace in Split, a Roman complex now brimming with cafes, plazas and boutiques
- Zagreb's galleries, churches and museums
- Untouched forests and coves on Cres Island

Top things to do
- Sail, catch a ferry, or paddle a canoe along the sublime Adriatic coast
- Hunt truffles, sample the finest olive oil, or just enjoy a slow-food banquet in Istria, Croatia's gastronomy capital
- Get your gear off and plunge into the sea in your birthday suit on the islands off Hvar, and plenty of other spots besides
- Nod and smile at passers-by during the *corso*, the communal, early-evening promenade that happens in most Croatian towns

Getting under the skin
Read: *Our Man in Iraq*, the Croatian bestseller by Robert Perišić; or *Farewell, Cowboy*, a small-town and family drama by Olja Savičević Ivančević
Listen: to traditional *tamburica* (lute) music or *klapa* (Dalmatian a cappella singing)
Watch: *Armin*, directed by Ognjen Sviličić, a poignant observation of father-son relationships on the road to Zagreb; or Vinko Brešan's hit black comedy *How the War Started on My Island*
Eat: *brodet* (slightly spicy seafood stew); *pašticada* (beef stewed in wine and spices); *paški sir*, a pungent sheep's cheese made on Pag Island
Drink: Ožujsko or Karlovačko, the two most popular beers; wine from Kvarner or Baranja

In a word
Dobar dan (Good afternoon)

Celebrate this
Croatia is the birthplace of its own version of naive art, a distinct style of 20th-century painting that features fantastical depictions of rural life. Some of the most important artists of the genre are displayed at the Croatian Museum of Naïve Art in Zagreb.

Random fact
The sea provides a soundtrack in the town of Zadar, where an open-air organ is powered by the movement of the waves.

1. Dubrovnik's old town has its origins in the 7th century and was restored in the late 1990s after war
2. Plitvice Lakes National Park features 16 interlinked lakes
3. Looking out over Zagreb, one of Europe's top cities for nightlife with bars, music and dancing
4. Discover vineyards and speciality foods in hilltop towns like Motovun, deep in Istria

CAPITAL HAVANA // POPULATION 11.1 MILLION // AREA 110,860 SQ KM // OFFICIAL LANGUAGE SPANISH

Cuba

Cuba has been at the centre of intrigue for centuries, from the late 1800s when the US flexed its colonial muscles to its iconic revolution in 1959. Years of isolation have preserved colonial architecture from destruction even as it crumbles, but change is well and truly afoot. Creative entrepreneurs are revolutionising the food, accommodation and arts scene – things unheard of under former president Fidel Castro. Its raucous stew of African, Caribbean and Latin culture lures fans from around the globe.

Best time to visit
November to May, to avoid the heat and hurricanes

Top things to see
- Havana, the steaming, bubbling center of the nation
- The quieter side of the country in rural Valle de Viñales, with its limestone cliffs, tobacco fields and primordial forest
- Santiago de Cuba, an often-overlooked city with rich traditions
- Las Parrandas de Remedios (Christmas festival), year-round street-party nights, and languid beach days

Top things to do
- Spend the night in a *casa particular* (private homestay)
- Learn how to roll one of the country's fabled cigars at the tobacco plantations in Viñales
- Stroll the impromptu street festival that is Havana's waterfront Malecón
- Cheer on your team of choice at a baseball game rich with talent
- Experience some of the best diving in the world, spying giant lobsters and manatees, at Punta Frances on Isla de la Juventud, an unspoiled wonderland rich with life

Getting under the skin
Read: *Everyone Leaves* by Wendy Guerra about coming of age in post-revolution Cuba ; or *Enduring Cuba* by Zoe Bran for an insight into the fascinating country
Listen: to Cuban son and Cuban takes on reggaeton, the combo of hip hop, dance-hall and reggae
Watch: Everyone's favorite, the hit Havana comedy *Fresa y Chocolate*; or the music documentary *The Buena Vista Social Club*
Eat: home-cooked ajiaco stew, featuring potatoes, meat, plantains, corn and old beer
Drink: a minty, sweet rum mojito

In a word
Queué bolá? (What's up?)

Celebrate this
Cuba turns out some of the most inventive artists in the world, many of them women: from artisanal soap to recycled paper products to stained glass there is much creativity to enjoy on this island.

Random fact
Baseball was introduced by American dock workers in the late 1800s and remains the national sport; games at stadiums across the nation are, like the country itself, passionate, raucous affairs.

1. A spectacular restoration of El Capitolio, or the National Capitol Building

2. Harvesting tobacco leaves for cigars in the Valle de Viñales

3. The town of Trinidad in Sancti Spiritus Province has some fine colonial architecture

Best time to visit
April to May and September to October, avoiding the baking heat of summer

Top things to see
- Frescoes of sword-wielding angels in the Byzantine churches at Pedoulas and Kakopetria
- Amazing mosaics in the Greco-Roman cities of Salamis, Kourion and Pafos
- Mile after mile of untouched golden sand in the remote Karpas (Kırpaşa) Peninsula
- Ten thousand years of treasures in the museums of Nicosia (Lefkosia)
- Echoes of 1975 in the maze-like backstreets of Nicosia, where roads end abruptly at UN sentry posts

Top things to do
- Sip mountain wine in the pine-scented Troödos Mountains
- Feast on a Cypriot meze in the crescent-shaped harbour at Kyrenia (Girne)
- Gaze in awe at the Frankish churches, Ottoman mosques and Byzantine city walls of Nicosia
- Swim with turtles in the Karpas and Akamas Peninsulas
- Storm the Crusader castles at St Hilarion, Kolossi and Kantara

Getting under the skin
Read: *Journey into Cyprus* by Colin Thubron or *Bitter Lemons of Cyprus* by Lawrence Durrell for evocative descriptions of pre-partition Cyprus
Listen: to Pelagia Kyriakou's *Paralimnitika* recordings, a superb collection of Cypriot folk songs
Watch: *The Story of the Green Line*, by Panikos Chrysanthou, or *The Slaughter of the Cock*, by Andreas Pantzis, for insights into the Turkish invasion of 1974 and its aftermath
Eat: the classic Cypriot *mezedes* – a pick-and-mix of savoury bites, from fried halloumi cheese to *sheftalia* (skinless pork sausages)
Drink: the local firewater, distilled from fermented grape skins – Greek Cypriots call it *zivania*, Turkish Cypriots call it *rakı*

In a word
Avrio/yarhun (Tomorrow – the best time to do anything on this laid-back island)

Celebrate this
Settled by Phoenicians, Greeks, Romans, Crusaders and Ottomans, to name just a few, Cyprus has plenty to offer history buffs, with many ancient sites on both sides of the island.

Random fact
Almost 3% of the island is officially part of the UK thanks to the sovereign army bases at Akrotiri and Dhekelia.

1. The Orthodox Church in Pedoulas, high in the cool Troödos mountains

2. Street life in Phaneromeni, the old quarter of Nicosia

3. Popular Konnos Bay in party-central Ayia Napa

C   **CAPITAL** NICOSIA (LEFKOSIA/LEFKOŞA) // **POPULATION** 1.26 MILLION // **AREA** 9251 SQ KM // **OFFICIAL LANGUAGES** GREEK & TURKISH

Cyprus

Basking in balmy seas at the eastern end of the Mediterranean, Cyprus is an island with two distinct sides. After the Turkish invasion of 1974, this sleepy isle was split into Greek and Turkish entities and, despite attempts at rapprochement, the two still coexist like estranged siblings in different wings of the family home. The southern, Greek-speaking half is a vision of the package-holiday Mediterranean, while the politically isolated Turkish part to the north feels like the year 1975 just popped outside for a glass of *rakı* and never came back. Yet the two sides of Cyprus have more in common than they might admit: fine beaches and sea turtles, pine-forested mountains, terracotta-tiled villages and more historic ruins than you can shake an amphora at.

1. The historic centre of Telč is a Unesco World Heritage Site with Renaissance houses around a triangular market square

2. A view of the Vltava River from the Máj lookout near the village of Teletín

3. The city of Český Krumlov in Southern Bohemia has a 13th-century castle

C CAPITAL PRAGUE // **POPULATION** 10.7 MILLION // **AREA** 78,867 SQ KM // **OFFICIAL LANGUAGE** CZECH

Czech Republic

There's more to the Czech Republic (also known as Czechia since 2016) than Prague, but what a capital city! From the cobbled streets of Staré Město to the architectural majesty of the castle district, Prague could have been plucked from a medieval fairy tale, and even the constant tide of tourists doesn't diminish the magic. Get your Prague fix, then escape to the elegant Renaissance and baroque cities of Olomouc, Český Krumlov and Telč to marvel at more artistic and cultural heights. And be sure to make time for the famous local beer; Czechia was home to the original Bohemians, and even decades after the fall of communism, it feels like the victory party is still going on.

Best time to visit
April to June, or stunning Christmas time

Top things to see
- Centuries of history and legend in Prague Castle
- The Renaissance facades and Gothic arcades of the old town square in Telč
- Sedlec Ossuary's creepy 'bone chapel', assembled from thousands of human skeletons
- The charm of Olomouc, with its astronomical clock and religious architecture
- Elegant Český Krumlov on the serpentine bends of the Vltava River

Top things to do
- Stride out across Prague's iconic 14th-century Charles Bridge
- Savour a hoppy ale in Plzeň, home of the original pilsner, and České Budějovice
- Ponder the sad fate of European Jewish communities in the cemetery of Josefov
- Explore the forested hills of Šumava on foot or bike
- Take a treatment and mix with B-list celebrities in the elegant spas of Karlovy Vary

Getting under the skin
Read: *The Book of Laughter and Forgetting* by Milan Kundera on communist Czechoslovakia
Listen: to Antonín Dvořák's *Slavonic Dances*, or his religious masterpiece *Stabat Mater*
Watch: *Kolya*, an Academy Award–winning tale set during the Velvet Revolution
Eat: *knedlícky* (dumplings) and *svíčková na smetaně* (roast beef with sour cream and cranberries)
Drink: *pivo* (beer) at countless breweries and pubs

In a word
Dobrý den (Good day)

Celebrate this
Prague-based artist David Černý is famous for public sculpture and installations that both amuse and make points about modern life and national myth.

Random fact
The sugar cube was invented here.

D CAPITAL COPENHAGEN // POPULATION 5.8 MILLION // AREA 43,094 SQ KM // OFFICIAL LANGUAGE DANISH

Denmark

Neat, easily navigable and just plain nice, Denmark's big draw is its user-friendliness. Its history is impeccably preserved, as numerous castles and medieval towns testify, but it's also state of the art with a slick approach to design, forward-looking social developments, smooth-running public transport and world-class restaurants. This blend fosters the nation's distinctive style and wins it a regular chart-topping place among the happiest nations on Earth. You won't have to search hard to find some much-prized *hygge*, a sense of cosiness, camaraderie and contentment. But *hygge* comes with a dash of introspectiveness, and perhaps this has given rise to the country's colourful legacy of fairy tales: with all those fortresses about, you won't need to wander long to run into a scene straight out of Hans Christian Andersen.

Best time to visit
May to September

Top things to see
- Cobbled streets in Ribe, Denmark's oldest town
- Danish style personified in the streets, stores and restaurants of Copenhagen
- Viking history up close and personal at Roskilde's Viking Ship Museum
- The gleaming white chalk cliffs of Møns Klint

Top things to do
- Escape to the beaches and bike trails on the island of Bornholm
- Stand with one foot in the Skagerrak (North Sea), the other in the Kattegat (Baltic Sea), at postcard-pretty Skagen
- Join 125,000 others at the Roskilde Festival, northern Europe's biggest music event
- Dine in one of Copenhagen's culinary hot spots
- Go kite-surfing at Denmark's 'Cold Hawaii', Klitmøller

Getting under the skin
Read: *Copenhagen Food: Stories, Traditions and Recipes*, in which Trine Hahnemann gives expert insights on the capital's gastronomy; or for a change of pace, Hans Christian Andersen's fairy tales
Listen: to Carl Nielsen for classical Danish music; or, for something completely different, Metallica, heavy metal pioneers co-founded by Dane Lars Ulrich
Watch: Academy Award winner *Babette's Feast*; anything directed by Lars von Trier; or gritty detective drama series *Broen* (The Bridge)
Eat: *smørrebrød* (open sandwich); *frikadeller* (Danish meatballs); *sild* (pickled herring); and of course Danish pastries, known locally as *wienerbrød* (Vienna bread)
Drink: *øl* (beer – the biggies are Carlsberg and Tuborg); or *akvavit* (schnapps)

In a word
Det var hyggeligt! (That was cosy!)

Celebrate this
From humble origins producing wooden toys in the 1930s to modern global phenomenon, Lego is, along with Vikings, Denmark's most famous export.

Random fact
Danes really do have an extraordinary inventive streak, bringing to the world many innovative creations including the loudspeaker, dry batteries and Google Maps.

3

4

1. Danish fairy-tale writer Hans Christian Andersen used to live in waterfront Nyhavn

2. Gjógv village on Eysturoy, one of the Faroe Islands in the North Atlantic that are a Danish autonomous territory

3. Buying baked pastries in Østerbro, Copenhagen, which is a city that values great food

4. The rainbow walkway by artist Olafur Eliasson atop the ARoS Aarhus Kunstmuseum

CAPITAL DJIBOUTI CITY // **POPULATION** 921,804 // **AREA** 23,200 SQ KM // **OFFICIAL LANGUAGES** ARABIC & FRENCH

Djibouti

Straddling the meeting point of three diverging tectonic plates, this tiny nation is one of Earth's geological marvels: fumaroles spew steam from the planet's insides; magma seethes beneath ever-thinning crust; dramatic, lunar-like deserts slowly sink; and luminous salt crystals precipitate on lakeshores well below sea level. For the visitor, this otherworldly spectacle is only enhanced by the fact that Djibouti's waters host some of the world's best diving. The culture – dominated by Somali and Afar peoples, and peppered with influences from Arabia, India and Europe – is equally intoxicating. Located on the Horn of Africa, Djibouti is of huge strategic importance, and hosts military bases of numerous nations: USA, China, France, Italy, Saudi Arabia and even Japan (its first overseas base since WWII).

Best time to visit
November to mid-April, when temperatures are tolerable

Top things to see
- Whale sharks and armadas of manta rays swimming past you in the Bay of Ghoubbet
- Afar tribesmen gathering gleaming salt crystals from the blinding floor of Lac Assal
- The ancient juniper forests in the national park of Fôret du Day, one of Djibouti's rare green splashes
- French legionnaires rubbing shoulders with traditionally robed tribesmen in the streets of Djibouti City
- Superb shipwrecks nestled on the seafloor of the Gulf of Tadjoura

Top things to do
- Visit Mars in the otherworldly landscape that is Lac Abbé
- Unwind on Plage des Sables Blancs, Djibouti's most beautiful beach
- Take a guided walk up into the cool climes of the Goda Mountains
- Stand on the 'bridge of lava', perhaps the thinnest piece of the earth's crust
- Follow in the footsteps of Afar nomads along the ancient salt route

Getting under the skin
Read: *Le Pays sans Ombre*, a series of short stories by Djiboutian Abdourahman Waberi – non-French speakers can pick up Jeanne Garane's English translation, *The Land Without Shadows*
Listen: to Groupe RTD's 2020 album *The Dancing Devils of Djibouti* – it's the nation's first-ever globally released album
Watch: *Total Eclipse*, a film – partly shot in Djibouti – that follows the tumultuous life of French poet Arthur Rimbaud
Eat: *cabri farci* (stuffed kid) roasted on a spit; or *poisson yéménite* (fish suppers served in newspaper)
Drink: black coffee; or tea with lemon

In a word
Tasharrafna (Pleased to meet you)

Celebrate this
The elephant shrew species – neither an elephant nor a shew – was believed lost, but after 50 years of obscurity it was rediscovered in Djibouti in 2020.

Random fact
At 153m below sea level, Lac Assal is the lowest point on the continent of Africa.

1. The mineral spires of Lake Abbe were formed aeons ago when volcanic gases bubbled through water that was 30m deeper than it is today; Afar people herd their livestock through this barren landscape

2. Local boats ply the Gulf of Aden

3. Beneath the waves, whale sharks can be found in the waters around Djibouti

1. Looking towards Portsmouth and Prince Rupert Bay from Morne Espagnol on the northwest coast

2. Fort Shirley in Cabrits National Park was built by the British (with enslaved labour) to defend against French raids in the 18th century; it has been carefully restored

3. Taking a dip in Trafalgar Falls near Roseau

(D) CAPITAL ROSEAU // POPULATION 73,286 // AREA 754 SQ KM // OFFICIAL LANGUAGE ENGLISH

Dominica

Little-known Dominica is one of the few remaining Caribbean islands where mass-market tourism has not yet made its mark, and as such it's an incredible secret that will no doubt soon be out. Wedged between the two modern, French-speaking and very heavily developed islands of Martinique and Guadeloupe, English-speaking Dominica is a different world: the island is mountainous and shrouded in jungle. It is a walker's paradise and overflows with waterfalls, lakes and rushing rivers, while twisting dirt roads and scattered villages create a mood of isolation that's an escapist's fantasy.

Best time to visit
Mild trade winds keep Dominica beautifully pleasant year-round

Top things to see
- Sisserou parrots at Morne Diablotin National Park
- Morne Trois Pitons National Park, famed for its volcanic scenery, including Emerald Pool
- Kalinago Barana Autê, a recreated Indigenous village highlighting the only pre-Columbian people still living in the Caribbean
- Scotts Head, a minnow-sized fishing village on dramatic Soufriere Bay
- New Market in Roseau to see the locals gather and collect a picnic

Top things to do
- Swim through a narrow canyon at Ti Tou Gorge to a gushing waterfall
- Wonder at the natural fizz of the waters below the surface while diving at Champagne Beach
- Traverse deep canyons on a trek to Trafalgar Falls
- Percolate in the wonderfully named Boiling Lake, above a lava-filled crack in the earth's crust
- Serenely glide alongside whales on a sailing boat

Getting under the skin
Read: *Voyage in the Dark* by Jean Rhys; or Dominica's other noted novelist Phyllis Shand Alfrey
Listen: to African *soukous*, Louisiana zydeco and a variety of local bands at the annual World Creole Music Festival in Roseau
Watch: *The Sea Between Us,* about a group of people from Martinique who embark on a journey to establish connection with Dominica in the aftermath of Hurricane Maria in 2017
Eat: callaloo soup (made with dasheen leaves)
Drink: Sea moss nonalcoholic beverage, made from seaweed mixed with sugar and spices and sometimes evaporated milk

In a word
Irie (Hi, bye, cool)

Celebrate this:
Dominica is dedicated to reducing energy consumption in all business sectors, getting all its energy from renewable resources and even supplying its neighbours with renewable energy.

Random fact
Dominica's national bird, the sisserou parrot, is the largest of all the Amazon parrots and thrives in trees along the island's 200 rivers.

Best time to visit
December to July (to avoid the hurricane season)

Top things to see
- Santo Domingo's Zona Colonial, Spain's original stepping stone into the New World
- Humpback whales, which gather every January to March to mate and give birth at Península de Samaná
- The white sand and turquoise waters of Playa Rincón, one of the Caribbean's finest beaches
- Carnival, celebrated most raucously in Santo Domingo, Santiago and La Vega
- Damajagua, a cascade of 27 waterfalls tumbling into glorious limestone pools

Top things to do
- Dive beneath the waves to explore the coast's myriad reefs and wrecks
- Pick up the Atlantic breeze at the kitesurfing haven of Playa Cabarete
- Go for the home run at a baseball match, the DR's (other) national religion
- Put on your dancing shoes to the heady rhythms of the merengue clubs
- Buzz from the adrenaline rush of white-water rafting on the Río Yaque del Norte

Getting under the skin
Read: *In the Time of the Butterflies*, a novel by Julia Alvarez about life under the Trujillo regime; or *Dominicana*, by Angie Cruz, about a 15-year-old girl who marries a much older man and immigrates to New York so her family can eventually join her
Listen: to merengue legends Johnny Venture and Coco Band; or stars of *bachata* (guitar music based on bolero rhythms) Raulín Rodriquez and Juan Luís Guerra
Watch: *Qué León* (2018) a comedy of star-crossed lovers staring a Puerto Rican–Dominican pop artist and a well-known Dominican-American model
Eat: *la bandera* ('the flag') – red beans, white rice and green plantain with meat stew
Drink: beer – a cold glass of Presidente at a sidewalk bar is a quintessential Dominican experience

In a word
¡Que chulo! (Great!)

Celebrate this
Baseball in the DR is more than a national pastime, it is an integral part of the nation's social and cultural landscape – check out a game at Estadio Quisqueya in Santo Domingo where the quality of play is high, dancers perform to merengue beats on top of dug outs and everyone is decked out in team colours.

Random fact
The foundation stone of Santo Domingo's Catedral Primada de América was laid in 1514, making it the oldest cathedral in the Americas.

1. Packing beans at an organic coffee plantation in Jarabacoa

2. Flamingos take flight over the Laguna de Oviedo

3. Beautiful Bahía de las Águilas runs along the southwest coast

D   **CAPITAL** SANTO DOMINGO // **POPULATION** 10.3 MILLION // **AREA** 48,670 SQ KM // **OFFICIAL LANGUAGE** SPANISH

Dominican Republic

Dominicans will tell you that it's no accident their country is one of the most popular tourist destinations in the Caribbean. 'What's not to like?' they'd ask, pulling you out for a night soaked in rum and merengue dancing. It's a hard argument to dispute. Santo Domingo's Zona Colonial exudes romance, the coast is fringed with white beaches and scuba diving sites, while the more energetic can delve into the lush green interior to trek along mountain trails. In a country that lists baseball diamonds alongside churches as hallowed ground, you can expect fun to be taken seriously – why else insist on having not one but two annual carnival celebrations?

Best time to visit
May to December on the mainland, or January to April for the Galápagos

Top things to see
- The splendid colonial centres of Quito and Cuenca, Unesco World Heritage Sites with foundations dating back to the 16th century
- Ecuador's Amazon rainforest, a vast region of unsurpassed biodiversity
- The 5897m-high Volcán Cotopaxi, best seen from atop a horse or from the windows of a mountainside hacienda
- Cloud forest reserves like those around Mindo or Parque Nacional Sumaco-Galeras, in which hundreds of bird species have been recorded

Top things to do
- Go wildlife watching in the Galápagos Islands and see giant tortoises, marine iguanas and sea lions
- Hike the Andes amid the striking scenery of the Quilotoa Loop
- Browse for ponchos, hammocks and jewellery in Otavalo, one of Latin America's largest craft markets
- Take to the rails on the impressively revamped train between Quito in the Andes and coastal Guayaquil

Getting under the skin
Read: *Century of the Death of the Rose* by Jorge Carrera Andrade, profound verse from one of Latin America's most influential poets
Listen: to La Máquina Camaleön, a Quito band who mix folk, pop and psychedelia
Watch: Tania Hermida's *Que Tan Lejos?* (How Much Further?), a road movie about two young women on an unplanned journey of self-discovery in the highlands
Eat: *encocado* (Afro-Ecuadorian seafood stew cooked with coconut milk and spices)
Drink: *canelazo*, a warming drink of hot *aguardiente* (sugar-cane alcohol) served with cinnamon, sugar and citrus

In a word
Naturaleza (Nature)

Celebrate this
Ecuador boasts the world's first two Unesco World Heritage Sites: twelve sites were chosen in the initial round of nominations in 1978, and the Galápagos Islands and Quito were #1 and #2 in the prestigious list – which is now over 1200 strong.

Random fact
Despite being only a little larger than the UK, Ecuador is home to some 300 mammal species and over 1600 bird species – more than Europe and North America combined.

1. The Galápagos land iguana on Santiago Island was decimated by invasive animals

2. Workers on the Tren de la Libertad, which runs from Ibarra to Salinas

3. Looking over bird-filled cloud forest from Mashpi Lodge

E **CAPITAL** QUITO // **POPULATION** 16.9 MILLION // **AREA** 283,561 SQ KM // **OFFICIAL LANGUAGE** SPANISH

Ecuador & the Galápagos Islands

This Andean nation appears diminutive beside its larger neighbours, yet there are years' worth of diversions packed between its jungles, cloud forests, 5000m peaks and awe-inspiring tropical coastline, including the nature lover's paradise of the Galápagos Islands. Culturally, things are just as varied, with a dozen indigenous groups, a sizeable Afro-Ecuadorian population and more recent immigrants from Asia. A period of serious government spending (partly based on oil revenues) in the 2010s, meanwhile, transformed infrastructure and education. With highland towns renowned for textiles, fishing villages thrumming with cumbia, rivers boasting South America's best rafting and some stunningly situated jungle lodges, this compact, laid-back nation is one of South America's highlights.

Best time to visit
October to May to avoid the heat

Top things to see
- Pyramids of Giza, the last intact Ancient Wonder of the World
- Ancient Egypt's heartland, Luxor, from the Valley of the Kings to the sublime Karnak Temple
- Cairo, the clamorous 'Mother of the World', with mosques, mausoleums and the Egyptian Museum
- The Abu Simbel temple complex, arguably the most beautiful landmark from Ancient Egypt
- The White Desert's surreal rock formations south of the western oasis of Bahariya

Top things to do
- Watch the Nile's riverbanks in Aswan glow green and gold at sunset while onboard a felucca (traditional sailing boat)
- Dive into some of the world's richest aquatic spectacles in Ras Mohammed National Park
- Stroll Alexandria's Corniche to soak up its fading 19th-century grandeur
- Chill, figuratively and literally, on Dahab's beach and in its Red Sea waters

Getting under the skin
Read: Naguib Mahfouz's *The Cairo Trilogy*; Alaa Al Aswany's masterful *The Yacoubian Building*; and Peter Hessler's *The Buried: An Archaeology of the Egyptian Revolution*
Listen: to Umm Kolthum, forever Egypt's diva
Watch: Oscar-nominated documentary *The Square* by Jehane Noujaim; and the acclaimed *678* by Mohamed Diab, both windows into modern Egypt
Eat: *kushari* (noodles, rice, black lentils and dried onions, served with a fiery tomato sauce)
Drink: *karkadai* (hibiscus juice), mint tea and *ahwa* (Arabic coffee)

In a word
Inshallah (God willing)

Celebrate this
Hassan Ali Ghazaly, founder of the African Youth Bureau, was chosen as Person of the Year at the Young African Leaders Summit 2019.

Random fact
The average length of bandages used to wrap bodies in the mummification process was 1.6km.

1. Buying and selling in Cairo's souq

2. Construction at the Karnak temple complex started 4000 years ago

3. The Nile brought floods and fertile soils to Egypt, aiding agriculture; the felucca was the main means of transport along the river

4. Siwa oasis lies close to the Libyan border

PETER SEAWARD | LONELY PLANET // PETER SEAWARD | LONELY PLANET // MARK READ | LONELY PLANET // LENINGAD1975 | GETTY IMAGES

E **CAPITAL** CAIRO // **POPULATION** 104 MILLION // **AREA** 1,001,450 SQ KM // **OFFICIAL LANGUAGE** ARABIC

Egypt

Historically, geographically and culturally, Egypt is a colossus. Its past reads like a grand epic and the country is rich in signposts to this story, from the peerless glories of Ancient Egypt and the jewels of Islamic Cairo, to the recent cultural and political cracks from the Arab Spring. Egypt's landscape has the quality of a myth, from the Sahara and its mysterious Cave of Swimmers or the Nile, that great river of legend, to the rich marine life of the Red Sea coast. And the collision of Middle Eastern, African and Mediterranean worlds here has long proved to be an intoxicating and overwhelming mix. Dive into its squiggling souqs, donkey-filled backroads and chaotic city streets, then come up for air atop a vast desert dune – you won't be disappointed.

E

Best time to visit
November to March (the dry season)

Top things to see
- Hot springs, highland coffee farms and artists' workshops on the Ruta de las Flores
- The hours' quiet passage in Alegría, the country's mountain-top flower capital
- Rugged ridges and mountain grandeur in Parque Nacional El Imposible
- The bucolic highland town of San Ignacio, in the shadow of the nation's highest peak
- Dreamy white-sand beaches around La Libertad

Top things to do
- Sample marinated rabbit or grilled frog at colonial Juayúa's popular weekend food festival
- Strike up a conversation with an ex-guerrilla guide at the Museo de la Revolución Salvadoreña in Perquín
- Shop for *sorpresas* (intricate folk scenes carved in ceramic shells)
- Conquer the longest break in Central America at surfing mecca Punta Roca
- Hike up to the volcanic crater lake at Volcán Santa Ana

Getting under the skin
Read: the bold erotic poems of Claudia Lars; or *Salvador* by Joan Didion, about the early days of the civil war

Listen: to the Latin pop of Marito Rivera, reggae act Anastasio y los del Monte and melodic '60s rockers Los Vikings

Watch: *Salvador*, the story of a war correspondent directed by Oliver Stone, for Hollywood's insights into the civil war; *Romero*, with Raul Julia, a true story about the high price of opposing tyrannical leadership

Eat: *pupusas* (cornmeal pockets filled with farmer's cheese, refried beans or pork rinds)

Drink: *refrescos de ensalada* (juices with chunks of fresh fruit)

In a word
Que chivo (How cool)

Celebrate this
El Salvador is known as 'The Land of the Volcanoes' for good reason: there are 20 in this small corner of Central America, two of which (Izalco and Santa Ana) can be climbed in a single day.

Random fact
Over a third of El Salvadorans live and work abroad, sending US$5 billion home yearly in remittances to support their families.

1. El Salvador's economy has been built on coffee beans; introduced in the 1880s, high-quality coffee accounted for 90% of the country's exports by the 1920s

2. Looking over a neighbourhood of San Salvador

3. The lake inside Santa Ana stratovolcano, the nation's highest volcano at 2381m

E CAPITAL SAN SALVADOR // POPULATION 6.5 MILLION // AREA 21,041 SQ KM // OFFICIAL LANGUAGE SPANISH

El Salvador

El Salvador is strong coffee for the senses. The smallest country in Central America has suffered high crime rates and corruption for many years, and most travellers have kept their distance. But there is a bright side to this resilient nation: the frank talk of civil war survivors is juxtaposed with artists' carvings in the brightly painted towns of the Ruta de las Flores (Route of the Flowers), while recycled US school buses done up as psychedelic chariots buzz with music and conversation right across the country. More than anywhere else in Central America, this is a chance to get off the beaten path and see the region for what it is – in all its grittiness, gaudiness and, sometimes, just plain gorgeousness. And El Salvador has surprises stashed up its sleeves: countless volcanic mountains, swimming holes, exquisite handicrafts and a wild Pacific coast.

(E) CAPITAL LONDON // POPULATION 55.2 MILLION // AREA 130,373 SQ KM // OFFICIAL LANGUAGE ENGLISH

England

The British Empire spread English culture around the globe, but this is where it all started, from the cricket bats to the bowler hats. The largest of the four nations that make up the United Kingdom, England is famous for football and fine art, film makers and fashionistas, real ale and roast beef, live bands and dead playwrights, a nice cup of tea and the BBC, the Ministry of Sound and the Ministry of Silly Walks, Queen, the Queen, and 5000 years of history. And the jokes about dull, tasteless food? They're a thing of the past too, thanks to a wave of celebrated chefs and a population that's starting to care about what's on its plate.

Best time to visit
May to September

Top things to see
- In London, make time for St Paul's, Tate Modern, the British Museum and a bar crawl through Shoreditch
- A gig by the 'next big thing' at the Leadmill in Sheffield or the Kazimier in Liverpool
- Giddy views over the Lake District from the top of Scafell Pike
- A home game at Arsenal, Chelsea or Manchester United
- The gorgeous cathedrals at York, Durham and Lincoln

Top things to do
- Relive Spinal Tap's finest hour at Stonehenge, England's most iconic prehistoric monument
- Laze away a sunny afternoon drinking bitter beer at an English country pub
- Soak up ancient history in the Roman baths at Bath
- Go crazy for curry in Birmingham's 'Balti Triangle'
- Slurp on a stick of seaside rock on Brighton Pier

Getting under the skin
Read: too many options but start with Austen for veiled satire, Dickens for historical context, and Jeremy Paxman's *The English* for a contemporary assessment
Listen: to world-conquering rock and pop from The Beatles to The Artic Monkeys, and David Bowie to Adele; classical from Elgar and Vaughan Williams
Watch: *The Madness of King George*, directed by Nicholas Hytner; and *This is England*, directed by Shane Meadows, for contrasting explorations of English culture
Eat: fish and chips; curry; a full English breakfast; Sunday lunch with all the trimmings; sticky toffee pudding
Drink: tea; craft beer from a growing number of independent breweries

In a word
Cheers!

Celebrate this
Confounding the stereotype of the self-interested millionaire soccer star, footballer Marcus Rashford campaigned for free school meals for disadvantaged children in 2020 and then started a child reading initiative in 2021.

Random fact
In a long list of inventions, one surprising English innovation is said to be sparkling wine (don't tell the French).

1. Local crab delivery in Cornwall in the southwest of England
2. Browse the stalls of Maltby Street food market under railway arches in London
3. Sail the Norfolk Broads in the east of England
4. Haworth village in Yorkshire, where the Brontë sisters lived and wrote such novels as *Wuthering Heights*

Best time to visit
December to February for Bioko Island; May to September for Rio Muni

Top things to see
- Marine turtles climbing up the black sands at Ureca to lay their precious eggs
- Hâkâ, the lengthy sandbar that snakes into the dark depths of the Atlantic from Isla Corisco's palm-dotted shore
- Hippos, goliath frogs and turtles at home in the wilds of Reserva Natural de Rio Campo
- The tug of war between natural beauty and the oil industry off Bioko's coast
- Plaza de España and the Gothic Revival cathedral in the heart of Malabo

Top things to do
- Push through undergrowth in Monte Alen National Park's jungle while on a quest to see gorillas, chimpanzees and forest elephants
- Step into a pirogue and explore the wild fringes of the Estuario del Muni near Gabon's frontier
- Settle into the bleached sands on Isla Corisco's deserted beaches
- Trek through the fern-tree forests on the volcanic slopes near Moka

Getting under the skin
Read: *The Wonga Coup: Guns, Thugs and a Ruthless Determination to Create Mayhem in an Oil-Rich Corner of Africa* by Adam Roberts – the title says it all; *By Night the Mountain Burns* by Juan Tomás Ávila Laurel
Listen: to the lyrics of Eyi Moan Ndong and his chorus, accompanied by drums and the *mvet* (a cross between a zither and harp); the duo Hijas del Sol, who perform in both Bube and Spanish
Watch: the first feature film made in the country, 2014's award-winning *Where the Road Runs Out*, about a scientist who decides to discover his roots in Equatorial Guinea
Eat: seafood plucked from the ocean that very day
Drink: *osang* (local tea); or *malamba*, brewed from sugar cane

In a word
Mbôlo (Hello, in Fang)

Celebrate this
Tutu Alicante: a human rights lawyer and activist who founded the NGO EG Justice; he works to build lasting democratic reforms in Equatorial Guinea and to highlight rampant corruption in the government

Random fact
Since 2012 the government has been working to build the nation's new capital city; located in the interior of the Rio Muni mainland, and originally known as Oyala, the capital is formally called Cuidad de la Paz

1. The mandrill, native to Equatorial Guinea, is the world's largest monkey

2. The Basilica of the Immaculate Conception of the Virgin Mary in Mongomo

3. Rainforest meets the Atlantic Ocean on the 195km coast

E CAPITAL CUIDAD DE LA PAZ // POPULATION 836,178 // AREA 28,051 SQ KM // OFFICIAL LANGUAGES SPANISH & FRENCH

Equatorial Guinea

A country of two worlds, Equatorial Guinea is divided not only by sea but also by oil. Large reserves of black gold were discovered beneath the ocean's floor off the coast of Bioko Island in the mid-1990s, and the subsequent industrial development forever changed the island's landscape, economy and culture. Yet, parts of the island remain little changed, with fishing villages clinging to the coast, painted-faced primates and birds lurking in rainforests, and sea turtles nesting on its volcanic black-sand beaches. The country's mainland (Rio Muni), some 160km southeast of Bioko, has seen a fraction of the island's development, and much goes on here as it did centuries ago. Exploring its wild jungles, remote villages and pristine white-sand beaches are an authentic adventure into the Africa of old.

1. Asmara is packed with art deco architecture, such as this switchboard-inspired cinema by architect Mario Messina

2. The Dahlak Archipelago near Massawa is famous for its pearl fisheries

3. An Afar woman from central Asmara, wearing traditional beaded attire

E

CAPITAL ASMARA // **POPULATION** 6.1 MILLION // **AREA** 117,600 SQ KM // **OFFICIAL LANGUAGES** TIGRINYA, TIGRE, KUNAMA, AFAR, ARABIC, ENGLISH & OTHER CUSHITIC LANGUAGES

Eritrea

Etched into the Horn of Africa, the tiny nation of Eritrea is one of the world's most secretive countries. The repressive political climate over the past decade has warmed little, the only notes of thawing being the temporary opening of the border with long-time adversary Ethiopia in 2018 and the re-introduction of some tourism (a fraction of the country is accessible, with permits required to leave the capital). Yet Eritrea still offers travellers some serious rewards: Asmara's art deco architecture is dazzling, as are the numerous archaeological sites dotting the Abyssinian escarpments, peaks and plateaus. Despite the nine ethnic groups that make up the nation's diverse cultural fabric all facing economic and political hardships, they are always welcoming and a highlight to all visitors.

Best time to visit
October to May, when temperatures have cooled

Top things to see
- Inspired examples of art deco, expressionist, cubist and neoclassical Italian architecture in Asmara, the 'Piccolo Roma'
- Temple of Mariam Wakiro, one of the ancient archaeological ruins in Qohaito
- *Passeggiata*, an evening ritual when the people of Asmara take to the streets to stroll
- The view down over the Abyssinian landscape from the lofty monastery of Debre Bizen
- The tank graveyard on the outskirts of Asmara, hard evidence of past conflict's devastation

Top things to do
- Get cosy with corals while diving the depths of the Dahlak Archipelago
- Following the crowds walking from Keren to the weekly camel market outside town
- Inhale the aromas (and a macchiato or two) in the art deco cafes on Asmara's Harnet Ave
- Wend through the whitewashed buildings, porticoes and arcades of the island of Massawa

Getting under the skin
Read: *I Didn't Do it for You: How the World Used and Abused a Small African Nation* by Michela Wrong, an entertaining and angering portrayal of Eritrea
Listen: to *Greatest Hits* by Atewebrhan Segid, Eritrea's leading traditional musician and singer
Watch: *Heart of Fire*, a film based on the real life of a young female soldier in Eritrea's civil war; *Refugee: The Eritrea Exodus*, a documentary that tracks the plight of Eritrean refugees
Eat: *legamat*, deep-fried dough balls sold hot in newspaper cones
Drink: a piping-hot macchiato

In a word
Selam (Hello, in Tigrinya)

Celebrate this
One Day Seyoum: a youth-led organisation that campaigns about human rights violations against Eritreans, aids Eritrean refugees, and educates the wider public on the causes for which it fights.

Random fact
Eritrea has no official language; a multilingual state, its constitution confirms that all languages are equal in the country.

E CAPITAL TALLINN // POPULATION 1.2 MILLION // AREA 45,227 SQ KM // OFFICIAL LANGUAGE ESTONIAN

Estonia

Estonia has long stood apart from its neighbours, with its unique language and culture having survived all that history has thrown at it, whether it be occupation by Sweden, the Nazis or the Soviet Union. Now in its fourth decade of independence, this nation on the edge of the EU is at once fantastically modern and charmingly old fashioned. Soaking up Tallinn's long white nights and medieval history or exploring the country's coastline are joys to be savoured. Superb national parks get you back to nature, quaint villages evoke a sense of the past, and uplifting song festivals celebrate age-old traditions.

Best time to visit
May to September

Top things to see
- Tallinn's charming, photogenic medieval Old Town
- The bucolic splendour of Lahemaa National Park, with its manor houses and slices of deserted coastline
- Live music at the hugely popular midsummer Viljandi Folk Music Festival
- The futuristic, award-winning KUMU art museum in Tallinn's Kadriorg Park

Top things to do
- Island-hop along the west coast with a visit to both Saaremaa and Hiiumaa
- Go bog-shoe-walking and canoeing in the wetlands of Soomaa National Park
- Do a spot of cross-country skiing in the picturesque countryside around Otepää
- Get sand in your shorts at Pärnu, the country's summertime mecca
- Down a beer among students in the university town of Tartu

Getting under the skin
Read: *The Czar's Madman* by Jaan Kross, Estonia's most internationally acclaimed author; *Truth and Justice* by Anton Hansen Tammsaare, an insightful look at the country's 20th-century move to independence
Listen: to the austere but stunning music of contemporary composer Arvo Pärt
Watch: *Sügisball* (Autumn Ball), based on a 1979 novel portraying six residents of a drab high-rise apartment in Soviet-era Tallinn
Eat: *verivorst* (blood sausages) if you're feeling bloodthirsty; pork and potatoes (it's on every menu); smoked fish; fresh summer berries
Drink: Vana Tallinn – rum-based and spiced syrupy liqueur that is sweet and strong and best served in coffee, or over ice with milk

In a word
Terviseks! (Cheers!)

Celebrate this
Even under foreign rule, Estonia's fiendishly difficult language (it's one of only four EU tongues not to have an Indo-European root) has kept the spirit of an independent nation alive.

Random fact
Estonians invented the weird and wonderful sport of *kiiking*, whereby competitors stand on a swing and attempt to complete a 360-degree loop around the top bar.

1. Tallinn's Old Town is one of Europe's best preserved medieval centres

2. The watery wonderland of Lahemaa National Park is just a day trip away from Tallinn on the north coast

3. The university in the thriving town of Tartu is Estonia's national university

Best time to visit
March to November

Top things to see
- The annual Umhlanga (reed dance) in Eswatini's royal heartland
- Herds of zebra, wildebeest and antelopes from the back of horse in Mlilwane Wildlife Sanctuary
- Traditional crafts within the Malkerns Valley
- Preparations for the sacred Incwala ceremony around the royal *kraal* (enclosure) at Lobamba
- Brilliant orange flame trees and lavender jacarandas dotting the woodlands of the Ezulwini Valley

Top things to do
- Walk in the footsteps of giants in Mkhaya Game Reserve, perhaps the greatest place on the planet to witness rhinos (black and white) in the wild
- Zip line, hike or mountain bike through the rocky landscape of the Malolotja Nature Reserve
- Shoot the rapids while white-water rafting the Great Usutu River (Lusutfu River)
- Climb the massive granite dome of Sibebe Rock outside Mbabane for an endless vista
- Come face to face with lions, elephants and other iconic wildlife in Hlane Royal National Park

Getting under the skin
Read: *The Kingdom of Roses and Thorns* by Debra Liebenow, which follows five women facing challenges of modern life in Eswatini
Listen: to the songs and traditional rhythms of Bholoja's 'Swazi Soul' music; the hip hop of KrTC
Watch: the mix of performers at the annual Bushfire Festival, with theatre, music, poetry and more
Eat: *sishwala* (maize porridge)
Drink: *tjwala* (home-brewed beer), often served by the bucket

In a word
Yebo (Yes) – also an all-purpose greeting

Celebrate this
Bella Katamzi: the leader of Lutsango Regiment (a traditional troop of Eswatini women) who is working for the empowerment of women in Eswatini; she has been participating in the United Nations UN75 dialogues.

Random fact
Cows are revered in Eswatini, though they are traditionally only on the menu for men – it's believed that women will become too strong if they eat the meat, too vociferous if they eat the tongue, too smart if they eat the brains, and too likely to run away from their husbands if they eat the feet.

1. In Eswatini's reed dance ceremony, girls present the queen with cut reeds for her *kraal* (village of huts)

2. Elephant and vehicle stand-off at Hlane Royal Game Preserve

3. Constructing beehive huts at Mlilwane Wildlife Sanctuary

CAPITAL MBABANE (ADMINISTRATIVE), LOBAMBA (ROYAL & LEGISLATIVE) // **POPULATION** 1.1 MILLION // **AREA** 17,364 SQ KM //
OFFICIAL LANGUAGES SWATI & ENGLISH

Eswatini

Swaziland no more, the newly named Kingdom of Eswatini is a place of
great tradition. And its new title, bestowed by King Mswati III in 2018,
harkens back to just that. No longer labelled by its colonial past, Eswatini
(the land of the Swazis) is now rooted for good. This nation is Africa's
last absolute monarchy, and as such its ceremonies have changed little
in centuries past. No two are greater than the sacred Incwala ceremony
and the Umhlanga Dance Festival – both provide an incredible chance
to witness and to pay respect to Swazi culture. Other traditions are less
obvious, but they all underline the palpable warmth and enduring nature
of Swazis. Encountering Eswatini's mountainous landscape and varied
wildlife is also unforgettable, whether on foot, horseback, bicycle or 4WD.

Best time to visit
October to January, when the watered highlands are blooming marvellous

Top things to see
- The Mursi, Banna, Karo and other ethnic groups in the Omo Valley, an astounding cultural crossroad
- Troops of endemic geladas playing on Abyssinian precipices in the Simien Mountains
- The Ethiopian emperors' extraordinary castles in Gonder's Royal Enclosure
- A rare Ethiopian wolf darting across the Sanetti Plateau in Bale Mountains National Park

Top things to do
- Descend into Lalibela's medieval shadows to see the light: its 11 rock-hewn churches
- Wander the alleys of the ancient walled city of Harar
- Island hop across Lake Tana to visit remote, centuries-old monastic churches
- Emerge from Aksumite tombs and stare skyward at the ancient civilisation's stelae

Getting under the skin
Read: *Notes from the Hyena's Belly* by Nega Mezlekia, a coming-of-age story in the difficult decades surrounding the end of Haile Selassie's reign; *The Shadow King* by Maaza Mengiste, the incredible story of the Ethiopian women who went to war
Listen: to *The Very Best of the Éthiopiques*, a compilation of evocative Ethiopian jazz
Watch: *Difret*, about one girl's fight against abduction into forced marriage
Eat: *injera* (rubbery, sponge-like flatbread of national importance) laden with *berbere*-spiced beef and *gomen* (minced spinach)
Drink: a superb coffee (the bean is thought to have originated here)

In a word
Ishee (OK, hello, goodbye) – or with a smile it's a gesture of friendliness and goodwill

Celebrate this
Sara Mohammed – woman, mother, former model, and fashion designer – who is inspiring other women entrepreneurs with her company Next Design.

Random fact
When the Ethiopian People's Revolutionary Democratic Front tanks rolled into Addis Ababa in 1991, they were navigating with the map in Lonely Planet's *Africa on a Shoestring*.

1. A rare Ethiopian wolf in the central Bale Mountains National Park

2. Sulphur-infused Dallol lake in the Danakil Depression is one of the hottest places on Earth

3. Worshippers at Bet Giyorgis rock-hewn church in Lalibela, dedicated to St George

E // **CAPITAL** ADDIS ABABA // **POPULATION** 108 MILLION // **AREA** 1,104,300 SQ KM // **OFFICIAL LANGUAGES** AMHARIC

Ethiopia

Modern Ethiopia proudly blurs the line between past and present. Here, in Africa's oldest sovereign state – the only one that successfully fought off European colonial ambitions – Ethiopians embrace their ancient rituals and all the relics associated with their 2000-year-old civilisation. Ornate, rock-hewn churches dating back to the first millennium CE are not simply showpieces of past grandeur –they are active places of daily worship, hosting age-old ceremonies. These and other historical treasures – castles, tombs, stelae – pepper a mountainous landscape where animal species seen nowhere else on Earth play. Entering this unique world, with its own culture, language, script, calendar, timekeeping and wildlife, is as exciting as it is enlightening.

Best time to visit

May to October, when rainfall and humidity are lower

Top things to see

- Navala, Fiji's most picturesque traditional village
- Indo-Fijians walking across glowing coals at the South Indian fire-walking festival
- Taveuni's lush tropical flora and beautifully weathered mountains
- The immense windswept Sigatoka Sand Dunes on Viti Levu
- Orchids, walking tracks and lily ponds at the Garden of the Sleeping Giant

Top things to do

- Dive or snorkel with manta rays at Kadavu's Great Astrolabe Reef
- Island-hop through the Mamanuca and Yasawa islands, bask on beaches, snorkel and lap up local culture
- Swim through the ethereal Sawa-i-Lau caves made famous by Brooke Shields in *The Blue Lagoon*
- Birdwatch and trek through prehistoric rainforest to lofty Des Voeux Peak on Taveuni
- Surf world-class breaks – Cloudbreak, Wilkes Passage and Namotu Lefts – off the southern Mamanuca islands

Getting under the skin

Read: *Kava in the Blood* by Peter Thompson, an autobiography of a white Fijian imprisoned during the 1987 coup; or for something lighter, try *Sevens Heaven*, Ben Ryan's uplifting true tale of coaching the Fiji rugby team to Olympic glory
Listen: to the harmonies of choral music at Sunday service – a major part of Fijian life
Watch: *Pear Ta Ma'on Maf* (The Land Has Eyes), the tale of a girl's struggle with poverty and the strength she finds in her traditional mythology
Eat: a Fijian pit-cooked *lovo* (traditional feast) one day, Indian thali and roti the next, and Chinese specialities the day after
Drink: a bowl of traditional *yaqona* (kava) to numb your mind and your lips

In a word

Bula (Health, happiness, cheers, and 'bless you' if you sneeze)

Celebrate this

Rugby is a national obsession; one in every ten Fijians is a registered player. The Fiji Sevens is the only team in the world to win the Olympics, Sevens series and World Cup.

Random fact

Europeans adopted the name 'Fiji', which is actually the Tongan name of these islands; the inhabitants formerly called their home 'Viti'.

1. A Fijian paradise

2. Worshippers pour milk on a 'lingam' at Naag Mandir, a Hindu temple on Vanua Levu

3. Port city Lautoka is in the heart of Fiji's sugar-cane country and is known as 'Sugar City'

F CAPITAL SUVA // POPULATION 935,974 // AREA 18,300 SQ KM // OFFICIAL LANGUAGES FIJIAN, ENGLISH & FIJI HINDI

Fiji

Fiji is the embodiment of the South Pacific dream. Fall into a sun-induced coma on the beach, listen to palms rustling in the trade winds and dive into electric-blue seas. Inland there's a wonderland of traditional villages where, with your *sevu-sevu* (gift) of kava and cries of *'Bula!'*, you'll be greeted warmly. And how about some chai to wash down your taro? Indo-Fijians have introduced Hindu temples and curries, the Chinese community has made fried noodles omnipresent and Europeans have left their mark with colonial architecture and pizza. Decades of coups have slightly tainted the allure, but visitors tend to ignore politics as they bake on the beach.

1. In northern Finland, the aurora borealis tends to appear every other night in the darker months, less frequently in the south

2. The gorgeous National Library of Finland in Helsinki holds a copy of everything published in the country

3. Foraging for wild produce such as lingonberries is an important part of Finnish culture

F CAPITAL HELSINKI // POPULATION 5.6 MILLION // AREA 338,145 SQ KM // OFFICIAL LANGUAGES FINNISH & SWEDISH

Finland

Traditionally thrown in with Nordic neighbours Sweden, Norway and Denmark, Finland is Scandinavia's eccentric cousin from out of town: quirky, enigmatic but with a contagious and captivating self-confidence. So while you get the Nordic staples of cinnamon buns, hand-knit sweaters, cross-country skiing and ice-cool interior design, you also get Sami reindeer herders with hand-embroidered slippers listening to ear-splitting death metal and goth rock. This is, don't forget, the land that invented the sauna and the sport of wife-carrying. For both visitors and locals, the real appeal is the landscape – a vast, sparsely populated natural adventure playground, cloaked with silent forests and tranquil lakes. The trade-off is the cost of living, particularly when it comes to alcohol; it's easy to see why the Finns are the largest per capita consumers of milk and coffee.

Best time to visit
May to September to avoid the cold, or December to February for snow, Santa and the Northern Lights

Top things to see
- Beer terraces sprouting all over Helsinki at the first hint of summer
- Relics of Finland's tumultuous history at the fault-line between Russia and Sweden on the fortress island of Suomenlinna
- The mesmerising aurora borealis (Northern Lights)
- Opera performances at Olavinlinna Castle during the Savonlinna Opera Festival in July

Top things to do
- Join in one of the offbeat festivals like the wife-carrying or air-guitar world championships
- Sweat it out in the world's largest smoke sauna in Kuopio, interspersed with dips in the lake
- Get pulled through the snow by a team of huskies (or reindeer) in Lapland

Getting under the skin
Read: the Moomin children's books by Tove Jansson – stories of a family of lovable Nordic trolls
Listen: to the classical music of Jean Sibelius; or the *suomirock* of Lordi and Children of Bodom
Watch: Aki Kaurismäki's *Man Without a Past* or *Leningrad Cowboys Go America!*
Eat: fish (herring, whitefish); reindeer stew; Lapland cloudberries and lingonberries
Drink: coffee; or *salmiakkikossu* – a handmade spirit made from salt liquorice dissolved in vodka

In a word
Onko sauna lämmin? (Is your sauna warm?)

Celebrate this
Written records of them go back 900 years and 99% of Finns enjoy one every week: saunas, of course, invented here and beloved by naked locals.

Random fact
Finns are renowned for being quiet – there's an old joke that they invented text messaging so they wouldn't have to speak to each other.

F **CAPITAL** PARIS // **POPULATION** 67.85 MILLION // **AREA** 551,500 SQ KM // **OFFICIAL LANGUAGE** FRENCH

France

A conviction that they live in the best place on the planet is what makes most French tick. It's an assumption that accounts for a lot of stereotypes, positive and negative – arrogant, snooty, officious, opinionated, sexy and super-stylish – but when you consider that the country has produced some of the world's greatest philosophers, artists, musicians and literary figures you begin to wonder if the French have a point. Throw in a timeless land of deep-rooted tradition and modern innovation, a fabled feast of fine food and wine, a place which has unfaltering romance woven into every second footstep, a cinematic trip from opulent Renaissance chateaux to Parisian jazz bars to electric-blue seascapes and you might just end up agreeing that this really is the best place on Earth.

Best time to visit
Year-round

Top things to see
- Regenerated Marseille and the azure *calanques* dotting the nearby coast
- The mindblowing glacial panorama atop Aiguille du Midi (3842m), a cable-car ride from the resort of Chamonix in the French Alps
- Capital art in Paris: *Mona Lisa* at the Louvre, *The Kiss* at Musée du Rodin and cutting-edge contemporary at the Centre Pompidou
- The garden Monet painted at his home in Giverny
- Europe's highest sand dune (Dune du Pilat) overlooking views of amazing surf

Top things to do
- Taste Champagne in ancient cellars at Reims and Épernay in this Unesco World Heritage region
- Walk barefoot across kilometres of wave-rippled sand to Mont St-Michel
- Pedal through vineyards, cherry orchards and lavender fields in rural Provence
- Tuck into French gastronomic *art de vivre* in gourmet Bordeaux
- Motor the mythical corniches (coast roads) on the French Riviera

Getting under the skin
Read: Victor Hugo's phenomenal *Les Misérables*; *A Moveable Feast* by Hemingway on 1920s Paris; and Françoise Sagan's *Bonjour Tristesse* (Hello Sadness), published when she was just 18
Listen: to Serge Gainsbourg's breathless *Je T'Aime… Moi Non Plus* and feel your soul turn Francophile
Watch: *Le Fabuleux Destin d'Amélie Poulain* and imagine yourself a Parisian in Montmartre
Eat: crêpes in Brittany, Camembert cheese in Normandy, croissants in Paris and bouillabaisse soup in Marseille
Drink: cider in Normandy, pastis in Provence and well-aged red in Burgundy – and Champagne

In a word
Salut! (Hi!)

Celebrate this
Capturing people and places in paint is a centuries-old French speciality that has culturally enriched the world, from Claude's fantastical landscapes to Monet's fantastic *Water Lilies* series.

Random fact
Among other things, the French invented the digital calculator, hot-air balloons, Braille and margarine.

1. The public Plage du Midi in Cannes; the city is famous for its annual film festival

2. The gorgeous old walled town of St-Jean Pied de Port in southwest France is a popular starting point for would-be pilgrims on the Santiago de Compostela

3. Paris, city of beauty, culture, romance and some very fine bakeries

G CAPITAL LIBREVILLE // POPULATION 2.2 MILLION // AREA 267,667 SQ KM // OFFICIAL LANGUAGE FRENCH

Gabon

In the past couple of decades Gabon has been touted as Africa's last Eden. After all, this is where hippos surf, buffaloes bask on the beach and elephants roam vast slabs of equatorial rainforest. Yet it's now clear that not all is well. Despite the government doubling funding for its 13 national parks, which cover more than 10% of the nation's landmass, the once burgeoning ecotourism industry has failed to find its feet and poaching in the north has killed 80% of the elephant population in the remote Parc National de Minkébé. Throw in a hotly disputed election in 2016, a presidential stroke in 2018, and an attempted coup in 2019, and the peace and stability this nation was also known for seems to be in jeopardy.

Best time to visit
May to August (the dry season)

Top things to see
- Hippos surfing in the Atlantic swells breaking onto Parc National de Loango's shores
- A bird's-eye view of gorillas and forest elephants from Langoué Bai's observation platform in Parc National d'Ivindo
- Cirque de Léconi, the Bateke Plateau's spectacular red-rock abyss
- Giant sea turtles nesting on the beaches of Parc National de Mayumba
- Fire dancers lighting up the night in a traditional Bwiti initiation ceremony

Top things to do
- Track elephants, chimps, mandrills and lowland gorillas in the jungle of Réserve de la Lopé
- Bodysurf on Mayumba's deserted beach between humpback whale sightings
- Roll through dense jungles and dramatic landscapes on the Transgabonais railway
- Meander along the scenic Rive Droite (Right Bank) in Lambaréné to Albert Schweitzer's landmark hospital
- Hit the dance floors of Libreville's Quartier Louis when they kick off after midnight

Getting under the skin
Read: Michael Fay's Megatransect expedition reports from his 3000km, 15-month walk across Gabon and other parts of Central Africa
Listen: to anything by Pierre Akendengué, the 'father' of Gabonese music; or the *afro-zouk* and *soukous* songs of the late great Oliver N'Goma
Watch: *Boxing Libreville*, a documentary about a young boxer whose struggles mirror Gabon's issues with democracy
Eat: smoked fish with rice and *nyembwe* (a sauce of pulped palm nuts)
Drink: Beaufort – a local pale lager brewed in Libreville

In a word
Mbôlo (Hello, in Fang)

Celebrate this
Aida Touré: a multi-talented Gabonese artist, painter, Sufi poet and composer.

Random fact
High on nature – Gabon's forest elephants are particularly fond of iboga, a shrub known for its strong hallucinogenic properties.

1. Gabon's critically endangered forest elephants are known as gardeners of the forest, dispersing seeds and delivering dung
2. In the south, Mayumba's beaches are a favourite of surfers and turtles
3. Gabon's Ministry of Petroleum and Mines in Libreville; the country also exports uranium

G CAPITAL BANJUL // POPULATION 2.2 MILLION // AREA 11,300 SQ KM SQ KM // OFFICIAL LANGUAGE ENGLISH

Gambia, The

Caught in a permanent and loving embrace with the Gambia River, this sliver of a country (the smallest on the African continent) snakes its way almost 500km inland from the coast of West Africa. Its Atlantic beaches and surf have lured sun-starved Europeans for decades, but it is The Gambia's riverine environment that regularly welcomes far more visitors – namely the hundreds of migratory bird species on their way between Europe and Africa each year. This lush environment is also a haven for permanent residents, such as hippos, manatees, chimpanzees and several other primates. Exploring in The Gambia, whether its wilderness, thriving traditional music scene or ecotourism initiatives, is always full of surprises.

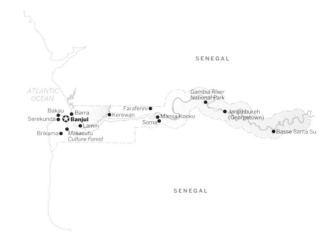

Best time to visit
November to April (dry season)

Top things to see
- World-class street art in rural villages as part of the Wide Open Walls project
- The 1200-year-old Wassu Stone Circles northwest of Janjanbureh
- Shafts of light cutting through the atmospheric smoke houses in the fishing village of Tanji
- Rare birds and giant crocodiles in Abuko Nature Reserve
- The Atlantic swells lapping on some of West Africa's finest beaches

Top things to do
- Journey by boat through the Chimpanzee Rehabilitation Project in Gambia River National Park, where great apes rule river islands
- Learn to play the traditional *kora* (21-stringed harp-like musical instrument) from a famous griot family in Brikama
- Interact with monkeys while walking in Bijilo Forest Park
- Take a pirogue ride through the mangroves of Makasutu Culture Forest
- Contemplate the nation's slavery past at the National Museum of Albreda

Getting under the skin
Read: *The Sun Will Soon Shine*, a short novel by Gambian author Sally Singhateh, which follows a young woman dealing with her past and her culture
Listen: to Bai Babu, one of The Gambia's top Mbalax artists; Sona Jobarteh's debut album *Fasiya*, which features *kora* instrumentals
Watch: *Kings of Gambia*, which follows a 10-piece band through The Gambia
Eat: *domodah* (peanut stew) or *benechin* (rice baked in a sauce of fish and vegetables)
Drink: *bissap* (hibiscus juice), *bouyi* (baobab juice) or JulBrew, the refreshing local beer

In a word
I be ñaading (Hello, in Mandinka)

Celebrate this
Jaha Dukureh: a Gambian activist who was nominated for the 2018 Nobel Peace Prize for her work helping to combat female genital mutilation

Random fact
The Gambia may be famous for its beaches, but it has the second-shortest coastline in Africa (80km), after Democratic Republic of Congo (37km)

1. The Banjul Central Mosque is the largest mosque in The Gambia

2. Boats are used for trade along The Gambia's coast

3. Visit the Chimpanzee Rehabilitation Project Camp where 100 rescued chimps live

❸

Best time to visit
April to October

Top things to see
- Tbilisi, a city of ramshackle elegance, balconied mansions, markets, a gallery-filled old town and thumping nightlife
- The hilltop silhouette of Tsminda Sameba Church in front of legendary Mt Kazbek
- The wild mountain provinces of Svaneti, Tusheti and Khevsureti, so remote that there are often no roads
- The imposing churches of Mtskheta, Georgia's spiritual heart since Christianity was introduced in 327 CE
- The wine region of Kakheti, including vaguely Tuscan-looking hilltop Sighnaghi

Top things to do
- Raise your glass with the locals for an alcohol-laden dinner and some polyphonic singing
- Head north to the mountains and take advantage of almost limitless opportunities for trekking between remote mountain hamlets
- Ponder the ascetic life in the monastery complex of Davit Gareja, located in the semi-desert near the border with Azerbaijan
- Sweat it up amid belle époque architecture and subtropical greenery in Batumi, Georgia's 'summer capital' on the steamy Black Sea coast

Getting under the skin
Read: *The Eighth Life* by Nino Haratishvili, a historical saga of 19th- and 20th-century Georgia
Listen: to extraordinary Georgian polyphonic singing – groups of singers often perform in restaurants for visitors, but you may also see a totally authentic and ad hoc performance in any mountain village
Watch: Levan Akin's *And Then We Danced*, featuring the first-ever gay storyline in Georgian cinema
Eat: *khachapuri* (cheese pie), or *khinkali* (spicy dumplings); delicious meats and stews, always cleverly spiced and flavoured with walnut paste
Drink: sublime Saperavi and Tsinandali wines, or, if you dare, *chacha*, the local firewater

In a word
Gaumarjos (Cheers)

Celebrate this
The Transcaucasian Trail is a project to establish a waymarked walking trail running the whole 700km length of the Caucasus, from the Black Sea to the Caspian Sea. It was initiated in 2015 by two former US Peace Corps volunteers in Georgia.

Random fact
Western Georgia, with the city of Kutaisi, is famous as the destination of Jason and the Argonauts in their search for the Golden Fleece.

1. La Fabrika is a multi-use venue in Tbilisi's hip Chugureti district

2. Tbilisi's Bridge of Peace crosses the Kura River from the Old Town

3. Kakheti is the centre of Georgia's ancient wine-making tradition

G CAPITAL TBILISI // POPULATION 4.9 MILLION // AREA 69,700 SQ KM // OFFICIAL LANGUAGE GEORGIAN

Georgia

At the heart of the mountainous Caucasus, Georgia is a land of centuries-old tradition and deep spirituality. In grand mountain scenery, wolves, bears and hyenas lurk, rivers race through steep gorges, and devout locals still cross themselves three times when they pass a church. Centred on its accessible and increasingly visited capital, Tbilisi – with excellent restaurants and wine bars, thriving club culture and a gorgeous old town – Georgia is nevertheless best appreciated by getting out to the mountains. Among the snow-capped peaks of the Caucasus visitors will experience wonderful hiking, ancient cave monasteries, and food and wine that will make you understand why many Georgians truly believe their land to have been blessed by God.

G CAPITAL BERLIN // POPULATION 80.2 MILLION // AREA 357,022 SQ KM // OFFICIAL LANGUAGE GERMAN

Germany

A powerhouse of Europe's economy, industry and innovation, Germany has given the world the printing press, the automobile, aspirin and audio digital technology, as well as seminal works by luminaries like Martin Luther, Albert Einstein, Karl Marx, Bach, Beethoven and the Brothers Grimm. At the continent's geographic heart, Germany's fairy-tale scenery spans the windswept maritime north, where white-sand-fringed islands scatter offshore, to dark, wooded forests, vineyard-ribboned river banks and the snow-capped Alps in the south. These diverse landscapes are sprinkled with half-timbered villages and towns, romantic castles, Roman relics, centuries-old breweries, and mighty cities home to palaces, museums, and traditional and cutting-edge drinking, dining and nightlife venues, along with a palpable sense of ancient and recent history.

Best time to visit
June to August (summer) for beer gardens at their best; September for beer festivals including Oktoberfest; late November to late December for magical *Weihnachtsmärkte* (Christmas markets)

Top things to see
- Berlin, which continues to reinvent itself three decades after the fall of the wall
- The soaring Gothic spires of Cologne's Unesco World Heritage–listed cathedral
- The inhuman horrors of Holocaust sites, such as Dachau and Buchenwald
- Dresden's Baroque Grünes Gewölbe, one of the world's richest treasure chests
- Munich with its Alpine accents, gregarious locals and fully realised good life

Top things to do
- Get lost in perfectly preserved small cities such as Bamberg or Weimar
- Float past vineyards and castles on a river cruise along the picturesque Rhine
- Find your favourite Bavarian beer hall or garden
- Hike the meadows and forests of the Alps
- Explore the tiny hamlets nestled amid the Harz Mountains

Getting under the skin
Read: Goethe's *Faust*, which tells of the classic deal with the devil, and the often-darker-than-you'd-expect *Grimms' Fairy Tales*
Listen: to classics by Bach or Beethoven, Berlin-style punk symbol Nina Hagen and Kraftwerk's groundbreaking '70s techno
Watch: cult classic *Run Lola Run* with vivid backdrops of Berlin's post-reunification city streets; *The Lives of Others*, a gripping look at life in the former East Germany
Eat: *Wurst* (sausages) such as *Bratwurst* and *Leberwurst*
Drink: Beers including *Dunkel* (dark), *Hell* (light), and *Weissbier* (wheat beer); Moselle and Rhine Rieslings

In a word
Wie gehts? (How's it going?)

Celebrate this
Germany is famous for cars but less known is the story of Bertha Benz (that name is a clue) who in August 1888 took her husband's petrol-engine vehicle on the first ever road trip, proving the viability of the car.

Random fact
Some 5000 varieties of beer are produced by Germany's 1300 breweries, including many small, local brewers; annual consumption averages 104 litres per person (behind the Czech Republic).

1. The hamlet of Sankt Roman is near Wolfach deep in the Black Forest; Wolfach is a spa town surrounded by trails and activities

2. Berlin's Museumsinsel houses 6000 years' worth of art, artefacts, sculpture and architecture from Europe and beyond in five grand museums

3. Münzgasse with Frauenkirche in the distance in Dresden; the East German city was largely rebuilt after WWII bombing

1. Looking out over the Gold Coast from the fomer Dutch Fort Batenstein; the name translates at 'profit fort'

2. Elmina Castle is one of the very first colonial European settlements in sub-Saharan Africa

3. Ghana's Asante people celebrate their ancestors at the Akwasidae Festival, which takes place every six weeks

G CAPITAL ACCRA // POPULATION 29.3 MILLION // AREA 238,533 SQ KM // OFFICIAL LANGUAGE ENGLISH

Ghana

Born from gold, and now fuelled by oil too, Ghana has long been a success story. The region's fertile lands and mineral wealth gave rise to several kingdoms, though it's the military prowess and fabulous wealth of the pre-colonial Asante (Ashanti) Empire that continues to capture the imagination of historians. Centuries later, after Europe's lengthy interest in gold and enslaved people waned, Ghana did what no other West African nation had dared to do – cast off the shackles of colonialism. Since then the nation has become a poster child for stability and democracy. Today, its rich cultures and proud African history combine with beautiful beaches, poignant slaving forts and large herds of elephants to capture the imagination of anyone who spends time here.

Best time to visit
November to March, for cooler and drier weather

Top things to see
- Elmina Castle, built in 1482 for trading gold, but which became an infamous colonial slaving fort
- Kejetia Market (West Africa's largest) in Kumasi, the ancient Asante capital
- Accra's busy markets, fine restaurants and sprinkling of historic buildings
- The porcupine-like mud-brick mosques of Wa

Top things to do
- Swim, hike and see waterfalls in the Volta Region
- Gain a chilling insight into the history of the slave trade at Cape Coast Castle
- Search out elephants in Mole National Park
- Learn how to play the *balafon* (West African xylophone) by the beach at Kokrobite
- Take a surfing lesson at Busua or laze on the beach at Akwidaa, Ghana's most beautiful sands

Getting under the skin
Read: *The Beautiful Ones Are Not Yet Born* by Ghanaian novelist Ayi Kwei Armah; *Changes: A Love Story* by Ama Ata Aidoo, about a young woman challenging the status quo in modern Accra
Listen: to Wiyaala, whose infectious tunes are often sung in Sissala and Waale dialects
Watch: *The Burial of Kojo*, a modern Ghanaian tale that is written and directed by Blitz Bazawule
Eat: *fufu* (pounded cassava or plantain) with a fiery sauce; *omo tuo* (rice balls in fish or meat soup)
Drink: *pito* (millet beer) in the north, palm wine in the south

In a word
Hani wodzo (Let's dance)

Celebrate this
Ama Ata Aidoo: a writer, poet and playwright who tackles social issues with creativity and honesty.

Random fact
At funerals, the dancing Nana Otafrija pallbearers have become meme sensations on YouTube.

(G) **CAPITAL** ATHENS // **POPULATION** 10.6 MILLION // **AREA** 131,957 SQ KM // **OFFICIAL LANGUAGE** GREEK

Greece

Odysseus lingered 10 years before coming home; Byron fell in love with the land and people; Lawrence Durrell wrote lyrically of island life: Greece seems to inspire all who come here. While it is commonly associated with blue seascapes and whitewashed villages, Greece, with a rugged Balkan hinterland, architecture from classical to modern periods and islands dotted across three seas, exhibits stunning diversity. There are olive groves, Chios' mastic villages, ancient Athens, crimson poppies every April, hirsute priests, old men sitting for hours over a single coffee, and ferries nudging into rickety piers. Many visitors come seeking sun and sea, but are smitten by the hospitality of the Greeks and find they are as captivated as all who went before.

Best time to visit
April to June, September (avoiding summer heat)

Top things to see
- The imposing white columns of the Parthenon, on a hill lording it over Athens
- Spectacular sunsets from Oia village at the northern tip of Santorini (Thira)
- The monasteries of Meteora astride rocky pillars on the plain of Thessaly
- Greek Easter in Corfu: priests in glorious vestments, candlelit midnight church services, sweet breads, coloured eggs
- The Knights' Quarter and Turkish relics in Rhodes' walled Old Town
- The underground museum in Vergina, burial place of Alexander the Great's son

Top things to do
- Hop on a ferry in Piraeus and cruise between Greek islands, each with their own character
- Imagine centaurs, satyrs or stray Greek gods as you trek up Mt Olympus
- Discover an undersea wonderland on a diving course in Ios
- Wander the mountain hamlets of the Zagorohoria, and trek the Vikos Gorge
- Mountain bike the gleaming coast of Halkidiki, mainland Greece's triple-pronged peninsula

Getting under the skin
Read: *Mythos* by Stephen Fry for an introduction to Greek gods and goddesses; and *Little Infamies* by Panos Karnezis for tales from the backwoods
Listen: to melancholy and passionate *rembetika*, sometimes called the Greek blues
Watch: the films of Theodoros Angelopoulos, including *Ulysses' Gaze* and *Eternity and a Day*
Eat: *saganaki* (fried cheese); *yemista* (stuffed peppers); *spanakopita* (spinach pastry); *soutzoukakia* (meatballs); grilled octopus; roast lamb; sticky *soutzouk loukoum* (candy rolled in nuts)
Drink: *ouzo* (grape brandy with anise); *retsina* (wine with resin); Greek coffee

In a word
Yamas (Cheers)

Celebrate this
Greece's 'love of wisdom' (aka philosophy) has a pedigree as old as its monuments and an impact that continues to this day: Socrates, Plato, Aristotle and co revolutionised thinking, making a huge contribution to Western civilisation.

Random fact
Around 500 BCE the poet Thespis is said to have improvised during a religious choral performance, thus becoming the first 'thespian', ie theatre performer.

1. Santorini's capital, Fira, is built on the edge of a caldera

2. Looking towards Lykavittos Hill over Athens from the Acropolis

3. Naoussa on Paros has evolved from a quiet fishing village into a stylish resort

4. The monasteries of Meteora perch on sandstone pillars

Best time to visit
The weather is warm throughout the year; January to April are the driest months

Top things to see
- St George's: a rainbow of Caribbean colours, ranged up the hillsides around the horseshoe-shaped Carenage Bay and topped by a fort
- Carriacou: the pint-sized sibling of the big island, with all the charms distilled down to an essence
- Petit Martinique: the smallest of the Grenada trio and a perfect paradise for seafarers
- La Sagesse Nature Centre, which preserves and promotes the local fauna and sits on a beach
- Diamond Chocolate Factory, a locally owned cooperative where you can follow the chocolate process from bean to bar

Top things to do
- Pound the crystalline sands at Grand Anse beach, Grenada's trademark beauty
- Dive into the underwater art gallery at Moliniere Bay, where coral and statues combine to create masterpieces
- Join the sea turtles at Anse La Roche, a hidden beach on isolated Carriacou
- Hike among mahogany trees and dew-dropping ferns in the Grand Etang rainforest, and spot some birds at the tranquil crater lake
- Build a sandcastle on Bathways Beach or soak up the secluded charms of next-door Lavera

Getting under the skin
Read: native Grenadian Jean Buffong's *Under the Silk Cotton Tree*
Listen: to reggae master David Emmanuel and Mighty Sparrow, the 'Calypso King of the World'
Watch: *Nothing Like Chocolate* – Grenada's brown-gold renaissance, as narrated by Susan Sarrandon
Eat: the national dish, 'oil down', which is spicy vegetables and meat cooked up in coconut milk
Drink: the nonalcoholic *mauby* (a bittersweet drink made from the bark of the rhamnaceous tree)

In a word
Sa ki fé'w? (What's happening? in Grenadian Creole)

Celebrate this
Grenada has some of the most advanced ecological laws in the region. The importation, sale and use of styrofoam containers is prohibited and the importation of plastic items, including shopping bags, plates and drinking straws is illegal. It is also reducing its dependency on fossil fuels and has committed to a 50% reduction by 2030.

Random fact
Grenada produces one-third of the world's nutmeg; the kernel is the odd-looking yellow blob on the left side of the Grenada flag.

1. Follow your nose in St George's spice market

2. Grand Anse is one of the island's most popular beaches

3. St George's was the capital of the island's spice trade, trading cinnamon, nutmeg and now cacao beans

G CAPITAL ST GEORGE'S // POPULATION 109,590 // AREA 344 SQ KM // OFFICIAL LANGUAGE ENGLISH

Grenada

Its nickname is Spice Island, and you can smell the nutmeg in the air on Grenada. But it also boasts luscious fruit bounty and a plethora of idyllic sandy beaches. The fact that this island chain is one of the most enchanting in the Caribbean is not up for debate. The three islands, Grenada, Petit Martinique and Carriacou all ooze charm and feel barely affected by mass package tourism. The hilly capital city has a charming waterfront and wide public beaches and the lush interior of the island is perfect for exploring. But don't miss the smaller islands, they are similarly idyllic, isolated and intoxicating.

Best time to visit
November to May (the dry season)

Top things to see
- Lost temples climbing above the jungle canopy at Tikal, the country's foremost Mayan ruin
- The gloriously pretty city of Antigua, with its Spanish-era convents, colonial ruins and backpacker bars
- Sunrise from atop Volcán Tajumulco, the highest point in Central America
- The tradition of Quema del Diablo, where lapping bonfires and fireworks psychically purge the year's trash
- The rugged road from Huehuetenango to Cobán, which teeters through highland coffee plantations

Top things to do
- Barter for rainbow-coloured Mayan textiles at the Chichicastenango market
- Paddle on or hike around the placid waters of Lake Atitlán, ringed by an honour guard of active volcanoes
- Finesse your bar talk by studying at a Spanish-language school in Antigua
- Soak in the cool emerald pools of Semuc Champey
- Board the breaks at Sipacate, the country's largely undiscovered surf capital

Getting under the skin
Read: *Hombres de Maíz* by the Nobel Prize–winning Miguel Ángel Asturias, which mixes magic realism with a critique of modern capitalism; the *Popol Vuh*, the founding myth of the K'iche' people
Listen: to *Guatemala: Celebrated Marimbas*; Latin pop superstar Ricardo Arjona
Watch: *Ixcanul* (Volcano): the debut of Guatemalan writer/director Jayro Bustamente, this 2015 feature about an arranged marriage within the indigenous Kaqchikel has won numerous international awards
Eat: a hearty Chapín (Guatemalan) breakfast of eggs, corn tortillas, beans, fried plantain and coffee
Drink: the velvety hot chocolate and Zacapa rum

In a word
Buena onda (Cool)

Celebrate this
Guatemala stretches higher than its neighbours, thanks to 4202m Volcán Tajumulco, the highest peak in Central America. But it also goes lower, with 340m-deep Atitlán the deepest lake and the Chiquibul Cave System (on the border with Belize) the deepest cave in the isthmus.

Random fact
Based on measurements of the passage of the stars, the ancient Maya were able to calculate the length of the solar year to within a few minutes.

1. The Arco de Santa Catalina in Antigua is a remnant of a 17th-century convent; the covered bridge enabled nuns to cross the street unseen

2. Chichicastenango holds popular twice-weekly markets but its cobbled streets are quiet at other times

3. A Maya woman wears traditional dress, including a *tocoyal* (headdress); the styles, patterns and colours used by each village are unique and represent each community's beliefs

4. Lake Petén Itzá, not far from Tikal, is surrounded by Maya ruins to explore

CAPITAL GUATEMALA CITY // **POPULATION** 17.2 MILLION // **AREA** 108,889 SQ KM // **OFFICIAL LANGUAGE** SPANISH

Guatemala

Guatemala is the heartland of the Maya people, whose rhythms flow seamlessly from the ancient world to the modern day. This is a country of beaches, waterfalls and coconuts, as well as highland towns where traditions are treasured and clothing is bright and elaborate. Political violence and instability have left scars, but the economy – and tourism in particular – has grown amid relative stability in the 25 years since the country's civil war ended. Visitors to Guatemala can indulge their best Indiana Jones fantasies: here colonial towns and coffee plantations cling to the slopes of active volcanoes and the ruins of vanished civilisations wait to be revealed by the next slash of your machete. Factor in a lively backpacker scene and some of the most colourful markets in the Americas and it's easy to see the appeal.

Best time to visit
November to February (the dry season – otherwise, Guinea is one of the wettest countries in the world)

Top things to see
- Colourful birdlife in Parc National du Haut Niger, one of West Africa's last dry-forest ecosystems
- The endless views from the former hill station of Dalaba
- Acrobats and contortionists from the Centre d'Art Acrobatique Keita Fodeba in Conakry
- The source of the Niger – the trickle that becomes one of Africa's greatest rivers

Top things to do
- Trek through the beautiful Fouta Djalon highlands to swim beneath waterfalls
- Encounter some of the region's last chimps while in the remote forest of Bossou
- Sip fresh coconut water on the beaches of Îles de Los, a short pirogue journey from the capital
- Wade into Conakry's frenetic live music scene
- Learn the *kora* (21-stringed instrument)

Getting under the skin
Read: *The African Child* (also called *The Dark Child*) by Camara Laye; *Warriors at Work: How Guinea Was Really Set Free* by Mustafah Dhada
Listen: to Ba Cissoko, who mix traditional *kora* sounds with modern influences
Watch: the ground-breaking *Dakan*, by Mohamed Camara, one of the first African movies to address homosexuality
Eat: *kulikuli* (peanut balls cooked with onion and cayenne pepper); grilled fish
Drink: *cafe noir*; or the beers Skol, Guiluxe and Castel

In a word
Bonne soirée (Have a good evening)

Celebrate this
Guinea's music star Mory Kanté, who passed away in 2020: known as the 'electronic griot', he revolutionised traditional West African music by blending Mandingo with electronic urban grooves.

Random fact
At least 22 West African rivers begin in the Guinean highlands, including the Niger, Senegal and Gambia Rivers.

1. Women sew new clothes in Conakry

2. Traditional accommodation beneath the Dame de Mali rock formation in the Fouta Djalon highlands

3. Stunning Sala waterfalls, one of many in northern Guinea

G **CAPITAL** CONAKRY // **POPULATION** 12.5 MILLION // **AREA** 245,857 SQ KM // **OFFICIAL LANGUAGE** FRENCH

Guinea

Although Guinea was always on the fringes of West Africa's early empires, its self-reliance and strength of mind was evident when it became the first nation in the region to defy France and claim independence. Yet it has been the thirst for control in governance that has held the nation back from its true potential, first with decades of dictatorial rule and now a worrying slide into ethnic politics since the first free and democratic elections of 2010. Behind the politics of power, Guinea's capital continues to be a vibrant beacon of world-class music and culture, and the little-developed mountains, plateaus and rainforests of the interior remain an enthralling (albeit challenging) place to explore.

G

Best time to visit
December to February, when it's dry and relatively cool

Top things to see
- Hippos eyes poking from the saltwater lagoons in Parque Nacional das Ilhas de Orango
- The white-sand beaches and lapping waters of Ilha de Rubane
- The crumbling Greek-style pillars of the abandoned town hall in Bolama
- Kere, the most beautiful island in the Arquipélago dos Bijagós

Top things to do
- Tango by candlelight in the cobbled streets of Bissau Velho
- Embrace island life within the villages of Ilha de Bubaque
- Follow footprints of buffaloes, elephants and chimpanzees in the dense forests of Parque Nacional de Cantanhez
- Dine on a fresh plate of oysters with hot lime dipping sauce after a day in the water at Quinhámel

Getting under the skin
Read: *The Ultimate Tragedy* by Abdulai Sila, a tale of love and politics in Guinea-Bissau
Listen: to the classic band Super Mama Djombo; and modern singers Dulce Maria Neves of Manecas Costa (*Paraiso di Gumbe*)
Watch: *My Voice*, a musical at the heart of Guinea-Bissau; *The Blue Eyes of Yonta*, by Flora Gomes, about dreams and revolution
Eat: *chabeu* (deep-fried fish served in a thick, palm-oil sauce with rice)
Drink: *cajeu* (a sickly sweet and dangerously strong cashew liquor)

In a word
Bom-dia (Good morning)

Celebrate this
At the 2019 World Athletics Championships, Guinea-Bissau's Braima Dabo gave up his medal chances in the 5000m to help an exhausted Aruban runner (Johnathan Busby) make it across the finish line.

Random fact
Citizens of Guinea-Bissau are known as Bissau-Guineans, not Guinea-Bissauans.

1. The carnival of Ilha de Bubaque in the Bijagós Islands is usually held the week before Lent

2. The long, sandy beach of Ilha de Rubane is one of the country's finest

3. Statues stand around the Portuguese-built 16th-century fort of Cacheu, once a centre of the slave trade

CAPITAL BISSAU // **POPULATION** 1.9 MILLION // **AREA** 36,125 SQ KM // **OFFICIAL LANGUAGE** PORTUGUESE

Guinea-Bissau

One of Africa's least-known corners, Guinea-Bissau is also one of its most beautiful and diverse. Wildlife-rich rainforests and decaying colonial-era towns dominate the mainland, and there are 23 different ethnic groups. But it's the offshore Arquipélago dos Bijagós, among the world's prettiest (and least-visited) island chains, that will really take your breath away. The peace and tranquillity of the archipelago's secluded coves and beaches stand in stark contrast to the turmoil that has blighted the country for decades. A brutal war of liberation was the precursor to Guinea-Bissau's extremely late independence from Portugal in 1974, and peace and prosperity have proved just as elusive ever since. Not that you'd know it, however, as Guinea-Bissau's people are just like their jokes and music – loud and tender.

1. A yellow-banded poison dart frog sends out a warning to predators in the Guyanan rainforest

2. During the rainy season, a vast volume of water shoots over Kaieteur Falls in an ancient, little-visited jungle

3. A giant river otter relaxes beside Rupununi River; they're a sociable species, living in large family groups

G CAPITAL GEORGETOWN // POPULATION 750,204 // AREA 214,969 SQ KM // OFFICIAL LANGUAGE ENGLISH

Guyana

Guyana may not be as well known as its larger neighbours, but with an eclectic cultural heritage and a geography that combines some of the world's most untouched rainforest with a distinctly Caribbean coast, it's a unique and intriguing nation. The present population is descended largely from enslaved Africans and indentured immigrants from India, while scattered indigenous settlements dot the interior. Guyana's rough-and-tumble capital, Georgetown, has a frontier aspect to it, though tensions tend to dissolve when the national cricket team takes the field. Beneath the headlines of corruption, power outages and economic trouble, Guyana is becoming an increasingly popular wildlife-watching and adventure destination. The pristine forests and incredible biodiversity within them are increasingly seen as the country's greatest assets.

Best time to visit
During the dry seasons: September to October and February to May

Top things to see
- Kaieteur Falls, the largest single drop waterfall in the world, located deep in the jungle and only accessible by small plane or three days' trek
- Wildlife, including giant anteaters, rare birds, jaguars and giant river otters, in the rainforests
- The Rupununi savannahs, sprinkled with indigenous villages, pockets of jungle and wildlife

Top things to do
- Stay at a community lodge on the Rupununi River before taking a dugout canoe to spot black caimans and anacondas
- Join *vaqueros* (cowboys) on a cattle drive in the remote Kanuku Mountains
- Travel by boat from Charity to Shell Beach, a sea turtle nesting site

Getting under the skin
Read: *Buxton Spice* by Guyanese writer Oonya Kempadoo, a coming-of-age tale set in the 1970s
Listen: British-Guyanese singer Eddy Grant (solo and with the Equals); Natural Black's reggae
Watch: *Guiana 1838*, Rohit Jagessar's story about the struggles against empire following the abolition of slavery and the ensuing arrival of labourers from India
Eat: pepperpot (a spicy stew cooked with different meats and a fermented juice made from cassava)
Drink: smooth, award-winning El Dorado rum

In a word
Howdy (How are you?)

Celebrate this
Guyana's interior is one of Earth's great wild spots: over 87% of the country is forested, and its jungles form part of the Guiana Shield, an area north of the Amazon with more than 3000 species of vertebrate.

Random fact
An astounding 500,000 Guyanese are estimated to live abroad – almost as many as in the country.

1

2

CAPITAL PORT-AU-PRINCE // **POPULATION** 11.2 MILLION // **AREA** 27,750 SQ KM // **OFFICIAL LANGUAGES** HAITIAN CREOLE & FRENCH

Haiti

Haiti's birth two centuries ago out of the only successful slave revolution in history has bequeathed it a proudly singular identity. Of all Caribbean countries, it is the closest to its African roots and has the richest tradition of visual arts, literature and music, as well as a deep spirituality that draws on its mixed Vodou and Christian heritage. Yet its recent history has been a difficult one, fraught with political instability and the 2010 earthquake that the country is still rebounding from. While the country's origin story makes others look like fairy tales, there is a certain joyful resilience that you'd do well to adopt upon visiting.

Best time to visit
December to July (to avoid hurricane season)

Top things to see
- The Citadelle fortress built to defend Haiti against Napoleonic invasion
- The fancy 'gingerbread' houses of Victorian-era Port-au-Prince
- Haiti's Museé du Panthéon National, housing King Christophe's suicide pistol and the anchor salvaged from Columbus' *Santa Maria*
- Congregations of Vodou faithful at the great celebrations at Saut d'Eau, Souvenance and Soukri
- The scrap metal turned cyberpunk Vodou sculpture of Port-au-Prince's Grand Rue artists' collective

Top things to do
- Catch a *tap tap* (local bus) across Port-au-Prince
- Party at Jacmel Carnival, one of the Caribbean's liveliest carnivals
- Pay your respects at a Vodou ceremony to get a new insight into this unfairly maligned religion
- Hike through the mountainous pine forests of Parc National La Viste
- Trek to the waterfalls and pools of Bassins Bleu

Getting under the skin
Read: *Haiti Noir 2*, a collection of work from Haitian writers edited by Edwidge Danticat; or post-earthquake reportage *The Big Truck That Went By: How the World Came to Save Haiti and Left Behind a Disaster* by Jonathan Katz
Listen: to Moonlight Benjamin – Vodou rock and Caribbean blues; Lakou Mizik – Haitian *kompa* (carnival sounds) and Vodou ritual music
Watch: *Ayiti Mon Amour* from Haitian film-maker Guetty Felin, a surrealist tale of three different people five years after the devastating earthquake; or Jonathan Demme's *The Argonomist*, about Haitian journalist and activist Jean Dominique
Eat: *griyo* and *banan peze* (fried pork and plantain) with *pikliz* (pickled slaw with chilli)
Drink: rum (preferably Barbancourt)

In a word
M pa pli mal (no worse than before) – the standard answer to 'How are you?'

Celebrate this
Complex and misunderstood, Vodou is a synthesis of traditional religions of West and Central Africa mixed with residual Taíno rituals and colonial Catholic iconography. Vodou played a large part in the struggle for independence and consequently was castigated in the US and Europe. It is now recognised as a national religion in Haiti alongside Christianity.

Random fact
In Haiti the spelling 'voodoo' is avoided due to lurid associations with Western popular culture.

1. Winter brings regular swells to Haiti

2. Jacmel in the south of Haiti has French colonial architecture that dates back 200 years

3. Citadelle Laferrière was commissioned by Henri Cristophe in 1820 as a fortress against the return of the French

Best time to visit
May to June for the festivals

Top things to see
- The extraordinary and intricate temples of Copán
- Rare whale sharks off the Caribbean coast from March to May and August to October
- The mellow, edge-of-the-jungle beach town of Trujillo
- Muddy jaguar prints in the wilderness in Río Plátano Biosphere Reserve
- Neighbourly goodwill at Guancasco, an annual Lenca ceremony promoting peace and friendship

Top things to do
- Get scuba certified, affordably, in the gemstone waters around Roatán
- Spot some of the 485 bird species that inhabit Lago de Yojoa
- Ride a dugout to the clear blue waters and white sands of Cayos Cochinos
- Hike past butterflies and monkeys in the spectacular cloud forest of Parque Nacional Celaque
- Glide down jungle rivers to find tapirs along the banks in La Moskitia

Getting under the skin
Read: *And the Sea Shall Hide Them* by William Jackson, an account of a murder off Utila, one of the Bay Islands; *The Soccer War* by Ryszard Kapuściński, about the 100-hour war between Honduras and El Salvador
Listen: Garífuna guitarist and singer Aurelio Martínez; the Garífuna-pop fusion of Banda Blanca
Watch: *El Espiritu de Mi Mama* (Spirit of My Mother), directed by Ali Allie, about a young Garífuna woman; Hispano Duron's *Morazán*, a tale of power and betrayal in the short-lived Federal Republic of Central America
Eat: fresh-from-the-sea *sopa de caracol* (conch soup) or *baleada* – Honduran flour tortillas stuffed with beans and cheese
Drink: ice-cold Port Royal or Salva Vida beer

In a word
Todo cheque (It's all cool)

Celebrate this
Honduras is home not just to the Maya calendar, but also to the oldest clock in the Americas: the machine, built in Moorish Spain in the 12th century, was transported here in 1620, and can be seen in Comayagua Cathedral.

Random fact
Honduras was the original banana republic – the American writer O Henry coined the phrase in the 1900s to describe the influence American banana companies wielded over the Honduran government.

1. A northern tamandua, an arboreal anteater and Honduras resident

2. Copán Ruinas is the handsome gateway to the Maya ruins nearby

3. Traditional dancers practise their moves in Tegucigalpa's Plaza Morazán

H CAPITAL TEGUCIGALPA // POPULATION 9.2 MILLION // AREA 112,090 SQ KM // OFFICIAL LANGUAGE SPANISH

Honduras

Honduras is the quiet child of Central America. It is blessed with abundant natural and historical riches, yet many people would struggle to mark it on a map. Dotted around this tropical nation are extravagant Maya ruins, atmospheric colonial cities, lush rainforests, reef-circled islands and the blissfully isolated inlets of the Mosquito Coast (La Moskitia), but Honduras has never seen large numbers of visitors, and when crime surged in the 2010s – a period that saw San Pedro Sula declared the world's murder capital – tourism slowed to a trickle. Things have improved since then, and the cities are worth a visit, although most visitors skip them in favour of the national parks, the Mayan pyramids and the languorous Bay Islands, where diving among whale sharks can provide the wildlife experience of a lifetime.

Best time to visit
April to June and September to October

Top things to see
- Fabulous views of Budapest's parliament building and the Danube River from the Fisherman's Bastion on Castle Hill
- The whip-cracking performances of *csikós* (cowboys) astride bareback horses in Hortobágy National Park
- Almond trees in blossom and ceramics, embroidery and other folk arts in Tihany
- The museums, mosques, beautifully preserved synagogue and Ottoman-era baths of Pécs
- The week-long Sziget music festival, one of Europe's biggest, on a leafy island in the Danube

Top things to do
- Plunge in for a hot soak and rub down in Budapest's steamy but elegant thermal baths
- Wet your whistle sampling feisty local wines in Eger's Valley of Beautiful Women
- Cruise the Danube on a ferry from Budapest to Szentendre, a former artists' colony
- Dip your toes in the northern shore of Lake Balaton, Hungary's freshwater 'riviera'
- Raise a glass in ruin bars, abandoned buildings given a decadent nightlife makeover, across Budapest and in Debrecen

Getting under the skin
Read: the postmodernist novels of 2015 Man Booker Prize–winner László Krasznahorkai, *Satantango* or *The Melancholy of Resistance*
Listen: to the *Hungarian Rhapsodies* of Franz Liszt or the haunting Hungarian folk music of Márta Sebestyén, as heard in the soundtrack to *The English Patient*
Watch: *Son of Saul*, the poignant debut by László Nemes that won the Oscar for Best Foreign Language Film in 2016
Eat: *paprikás csirke* (paprika chicken); or *gulyás* (goulash), full of beefy goodness
Drink: Tokaji Azsú, a sweeter-than-sweet dessert wine; Egri Bikavér, a full-bodied red known as 'bull's blood'; or homemade firewater *pálinka* (fruit brandy)

In a word
Egészségére (Cheers!)

Celebrate this
Budapest is the undisputed European capital of escape games, which are even claimed by some to be a Hungarian invention. Budapest has over a hundred escape rooms, often set in abandoned apartments. These mentally challenging games require lateral thinking and team spirit.

Random fact
The ballpoint pen was invented by Hungarian László Bíró.

1. A herd of Przewalski's horses in Hortobágy National Park

2. The art nouveau decor in Budapest's Gellért Baths is divine

3. Buda Castle was a royal palace from the 13th century until the 1918 revolution

CAPITAL BUDAPEST // **POPULATION** 9.8 MILLION // **AREA** 93,028 SQ KM // **OFFICIAL LANGUAGE** HUNGARIAN

Hungary

Hungary lies in the Carpathian Basin, slap-bang in the middle of Europe, and the Hungarians themselves will tell you that theirs is a Central (not Eastern) European nation – even though they trace their ancestry to beyond the Ural Mountains. It's a flat land dominated by the *puszta* (great plain), but Hungary's culture is anything but featureless. With a predilection for zesty paprika-infused cuisine and thermal baths (even during chilly winter days), traditions of horseback cowboy acrobatics and heel-clicking folk songs, and an inscrutable language, the Hungarians remain distinct from their neighbours. Hungary still pays homage to home-grown greats like Franz Liszt and painter Mihály Munkácsy, but today trendy coffee shops and ruin bars flourish alongside communist-era monuments and art nouveau architecture.

Best time to visit

May to September for long days and warm-ish weather; December to February to see Iceland at its iciest (and the Northern Lights)

Top things to see

- The mighty spurt of water from Geysir, the original hot-water spout after which all other geysers are named
- A breaching whale and curious dolphins on a whale-watching cruise from Húsavík
- The smouldering volcanic wastelands of Krafla
- The peaks and glaciers (and waterfalls and twisted birch woods) of Skaftafell
- Thousands of puffin chicks taking flight every August from Vestmannaeyjar

Top things to do

- Cavort with crowds of partygoers on the drunken Reykjavík *runtur*
- Breathe cool Icelandic air while bathing al fresco in volcano-heated water at the Blue Lagoon
- Explore the Westfjords on a kayaking trip under the midnight sun
- Cross the Arctic Circle with a visit to the island of Grímsey in Iceland's far north

Getting under the skin

Read: *Independent People* and other novels by Nobel Prize–winner Halldór Laxness; crime fiction from Arnaldur Indriðason; the sagas of the late 12th and 13th centuries
Listen: to the genre-defying works of Björk; blues/rock-inspired Kaleo; the ethereal sounds of Sigur Rós
Watch: *101 Reykjavík*, based on Hallgrimur Helgason's book of the same name – a painful, funny tale of a Reykjavík dropout's fling with his mother's lesbian lover
Eat: some challenging local dishes: *hákarl* (putrefied shark meat), or singed sheep's head complete with eyeballs
Drink: *brennivín* ('burnt wine'); schnapps made from potatoes and caraway seeds with the foreboding nickname *svarti dauði* (black death)

In a word

Skál! (Cheers!)

Celebrate this

The sturdy Icelandic horse is a treasured breed, developed from ponies brought by original Norse settlers. Strict rules govern them (exported horses are not allowed to return) and national celebrations recognise their cultural importance.

Random fact

Grímsey island is the only part of Iceland within the Arctic Circle – but that's changing. The Arctic Circle is shifting north, meaning the island, and thus the whole country, will be outside it in a few decades.

1. Looking out over Reykjavík, the world's most northerly capital, from the spectacular Hallgrímskirkja

2. Goðafoss lies on Iceland's ring road in the north of the country

3. The friendly Icelandic horses have more than 100 recognised colour combinations, each with its own name

4. Linger on the pale blue icebergs of Jökulsárlón, a lagoon just off the ring road in the southeast of Iceland

CAPITAL REYKJAVÍK // POPULATION 350,730 // AREA 103,000 SQ KM // OFFICIAL LANGUAGE ICELANDIC

Iceland

Iceland is a country in the making, a vast volcanic laboratory where mighty forces created by tectonic plates shape the land and shrink human onlookers to insignificant specks. See it in the gushing geysers, glooping mud pools, lava-spurting volcanoes and slow, grinding glaciers. Experience it first hand, bathing in turquoise-coloured hot springs, kayaking through a fjord, scanning the horizon for breaching whales or crunching across a dazzling-white icecap. Lower prices and an international recognition of its natural wonders have seen a huge influx of visitors in recent years, but away from the capital, Reykjavík, and the touristy Golden Circle it's still easy to drive on deserted roads, looking out for waterfalls, the unique Icelandic horses, traces of the country's Viking origins and those elusive Northern Lights.

CAPITAL NEW DELHI // **POPULATION** 1.3 BILLION // **AREA** 3,287,263 SQ KM // **OFFICIAL LANGUAGES** HINDI, ENGLISH, BENGALI, TELUGU, MARATHI, TAMIL, URDU, GUJARATI, MALAYALAM, KANNADA, ORIYA, PUNJABI, ASSAMESE, KASHMIRI, SINDHI & SANSKRIT

India

Crowned by the Himalaya, crossed by sacred rivers and coveted by empires from the Persians to the British Raj, India is vast and unfathomable, a kaleidoscope of cultures and the birthplace of at least three of the world's great religions. Countless civilisations have risen and fallen among its paddy fields, deserts, mountains and jungles, but India still endures. Asia's most colourful country has its problems – overcrowding, pollution, poverty, religious tensions, environmental degradation, grinding bureaucracy, and, most recently, the cruel toll of coronavirus – but this is all part of life's tapestry in this vivid and complex country. And for all the cacophony, there are captivating moments of serenity – dawn breaking over sacred rivers, sunset setting fire to snow-capped mountains, temple bells chiming like the music of the spheres.

Best time to visit
November to March in the plains; July to September for the Himalaya

Top things to see
- The white marble magnificence of the Taj Mahal
- Himalayan views and Raj-era oddities in Shimla and Darjeeling, the quintessential Indian hill stations
- The unbelievable tide of humanity crowding the banks of the sacred River Ganges in Varanasi
- An other-wordly landscape of mountains and monasteries in remote, breathless Ladakh
- The toppled temples of a vanished civilisation, scattered across the boulder-strewn badlands of Hampi

Top things to do
- Take an camel safari through the desert dunes of Rajasthan
- Scan the jungle for tigers in one of Madhya Pradesh's glorious national parks
- Kick back on the palm-brushed beaches of Goa
- Bend your body into shapes you never thought possible in Rishikesh, India's yoga capital
- Be overwhelmed by mountain majesty while trekking through the Indian Himalaya

Getting under the skin
Read: Salman Rushdie's *Midnight's Children*; Vikram Seth's *A Suitable Boy*; or Kiran Desai's *The Inheritance of Loss*
Listen: to the myriad *filmi* (movie soundtracks) recordings of Allah Rakha Rahman
Watch: Ramesh Sippy's Bollywood classic *Sholay*; Satyajit Ray's haunting *Pather Panchali*; or Zoya Akhtar's *Gully Boy*, to see how Indian cinema has evolved
Eat: delicious *thalis* (rice, curries, chapatis, pappadams and condiments, served on a metal platter or banana leaf)
Drink: *lassi* (sweet or salty yoghurt shakes); *chai* (sweet Indian tea) or Kingfisher beer

In a word
Jai hind! (Long live India!)

Celebrate this
At just nine years old, Licypriya Kangujam has delivered TED talks, led climate change rallies, and successfully lobbied the Indian parliament for new environmental laws, pushing environmentalism to the top of the political agenda.

Random fact
There is no such thing as curry in India – the South Indian word *kari* simply means 'fried' or 'sauce'.

1. Every evening at Dashashwamedh Ghat in Varanasi an elaborate and popular *ganga aarti* (river worship) ceremony with *puja* (prayers) is held

2. Near Nongriat in northeast India, the roots of rubber fig trees are guided to become living bridges

3. Pilgrims admire the Golden Temple of Amritsar, the spiritual home of Sikhism in Punjab

4. The Amber Fort, an ethereal example of Rajput architecture, rises from a rocky mountainside northeast of Jaipur in Rajasthan

Best time to visit
May to September, for dry skies (in most places)

Top things to see
- The geometry of the Buddhist stupa at Borobudur, a stone representation of heaven on Earth
- Shadow puppetry, batik dyeing and other ancient arts in Yogyakarta
- Death rituals that spill into the surreal in other-worldly Tana Toraja
- Rice terraces and tiered temples on touristy but sublime Bali
- Views over a primordial landscape of steaming volcanoes from the top of Gunung Bromo

Top things to do
- Soak up the barefoot beach vibe on the gorgeous Gili Islands
- Catch the perfect wave at Uluwatu, the Holy Grail of Bali surfing
- Come face-to-face with the world's largest lizard on Komodo and Rinca
- Experience the fascinating traditions of Dani tribal people on a trek to Papua's Baliem Valley
- Drop through a kaleidoscope of colour at world-class dive sites across the islands

Getting under the skin
Read: Ayu Utami's taboo-challenging *Saman*; or former political prisoner Pramoedya Ananta Toer's colonial-period saga, *This Earth of Mankind*
Listen: to the unmistakable sound of the *gamelan* – the traditional orchestra of Java, Bali and Lombok
Watch: the arthouse films of Garin Nugroho; or home-grown superhero romp *Gundala* by Joko Anwar
Eat: the ubiquitous *nasi goreng* (fried rice); or rich and spicy *rendang* (beef cooked slowly with roasted coconut and lemongrass)
Drink: Bintang beer; *kopi* (coffee)

In a word
Tidak apa-apa (No problem)

Celebrate this
Balinese sisters Melati and Isabel Wijsen founded the environmental organisation Bye Bye Plastic Bags at the ages of 12 and 10; their new project, Youthtopia, aims to train up a whole generation of environmental change-makers

Random fact
On Bali's annual Nyepi 'Day of Silence', the island – even the airport – shuts down for 24 hours

1. Spot an orangutan in the less frequented Gunung Leuser National Park of North Sumatra

2. Inside each of the 72 stupas at Borobudur in Central Java sits a Buddha statue

3. Hindu pilgrims bathe at Tirta Empul temple near Ubud on Bali, which is dedicated to the Hindu god Vishnu

4. Typical Tongkonan ancestral houses of the Torajan people in South Sulawesi; only nobles were permitted to build such elaborate homes

CAPITAL JAKARTA // POPULATION 267 MILLION // AREA 1,904,569 SQ KM // OFFICIAL LANGUAGE BAHASA INDONESIA

Indonesia

The world's largest archipelago stretches most of the way from the Malay Peninsula to Australia, spanning more than 17,000 islands, some tiny and some gargantuan. In fact, the number of islands changes regularly as earthquakes and volcanoes remodel the landscape. Fringed by blissful beaches and coral reefs, the islands are a vision in rainforest green, home to hundreds of rare plants and animals found nowhere else in the world. However, this astonishing biodiversity is under constant threat from development, and smoke from Indonesian forest fires casts a pall over large parts of Southeast Asia from April to October. For visitors, the variety of peoples and cultures on these scattered islands – from Islamic Java to Hindu Bali and the animist traditions of Borneo, Sulawesi and Papua – is as big an attraction as the islands' natural wonders.

Best time to visit
April to June and September to November

Top things to see
- The arched market arcades and beautiful mosques of Naqsh-e Jahan (Imam) Square in Esfahan
- Winding lanes and wind towers in the mud-brick old town of Yazd
- The tea terraces and hills surrounding Masuleh on the Caspian Sea littoral
- The ruins of Persepolis, an awe-inspiring reminder of the might of the ancient Persian Empire
- The domes and minarets of the Holy Shrine of Imam Reza in Mashhad

Top things to do
- Hike up to the Castles of the Assassins within the Alamut Valley
- Explore the desert scenery of the Dasht-e Kavir from its oases settlements of Garmeh or Mesr
- Accept an invitation to someone's home for dinner – you are sure to receive one – to experience first-hand Iranian hospitality
- Escape the smog and rumble of Tehran on the walking trails of Darband in the foothills of the Alborz Mountains

Getting under the skin
Read: *My Uncle Napoleon* by Iraj Pezeshkzad; and *Savushun* by Simin Daneshvar
Listen: to the sombre melodies of Persian epic poetry sung to traditional accompaniment
Watch: Oscar-winners *A Separation* and *The Salesman,* both directed by Asghar Farhadi; and *Gabbeh*, directed by Mohsen Makhmalbaf
Eat: *koresht e fesenjan* (chicken and walnut stew spiked with pomegranate molasses); and *zereshk polow* (rice and chicken with barberries)
Drink: *chay* (tea) at a *chaykhane* (traditional teahouse)

In a word
Khosh amadin (Welcome)

Celebrate this
Begun by Cyrus the Great, the Persian Empire at its height was the largest empire in the world, stretching from the Indus River in the east to the Balkans in the west.

Random fact
Iranians use a modified Arabic script, but their language, Farsi, is related to European languages.

1. The oldest ruins at Persepolis date back to 515 BCE

2. The interior of the Royal Mosque, just one of several stunning buildings on Unesco-protected Naqsh-e Jahan Square, and a high point of Persian architecture

3. Abyaneh village in central Iran is one of the oldest villages in Iran and has red mudbrick buildings

CAPITAL TEHRAN // **POPULATION** 84.9 MILLION // **AREA** 1,648,195 SQ KM // **OFFICIAL LANGUAGE** FARSI (PERSIAN)

Iran

Iran's history, recent and ancient, is nothing if not dramatic. From the ruins of the Persian Empire to the country's wealth of ornately tiled mosques and palaces, a glut of riches lie around every corner, all backdropped by a landscape which swoops from snow-capped mountains to desert plateau. Iran's modern face is just as diverse – cafe culture, snowboarding in the Alborz Mountains and hitting the capital's contemporary art galleries are all part of the lifestyle here. One of the Middle East's important political powers, Iran plays a decisive, often controversial, role in regional conflicts and regularly takes a bellicose stance with rivals. Meanwhile, internal tensions continue to play out between reformists and the country's conservative clerical establishment, sparking sporadic outbreaks of protest.

CAPITAL BAGHDAD // **POPULATION** 38.8 MILLION // **AREA** 438,317 SQ KM // **OFFICIAL LANGUAGES** ARABIC & KURDISH

Iraq

Incorporating a large swathe of Mesopotamia, Iraq was home to many of the great civilisations of the ancient world, with Baghdad as the Middle East's cultural powerhouse. Today, stability and security remains a distant dream for most Iraqis as a low-level insurgency rumbles along in the wake of the Iraq war. Many monuments of Iraq's golden age have been severely damaged by the conflict, though major historical markers such as the Lion of Babylon and the Great Ziggurat of Ur remain standing today. The most stable corner of the country is the semi-autonomous Iraqi Kurdistan region, which has a distinct and vibrant culture and contains some of the greenest mountain scenery in the Middle East.

Best time
April-May and September-October

Places of interest
- The citadel of Erbil, one of the oldest continuously inhabited cities on earth
- Hatra: most main monuments of this ancient trading city survive despite severe looting and destruction
- The architectural wonder of Taq Kasra (Arch of Ctesiphon), the world's largest brick vault
- Ur, with one of the world's best-preserved ziggurats
- Babylon: controversial reconstructions and damage sustained by stationed Coalition forces can't take away the significance of this site
- Lalish's Yazidi temples provide a glimpse into the culture of this Middle Eastern religion

Local activities
- For lush valley views and fresh air locals head to hilltop-perched Amadiya village
- The Hamilton Road in Iraqi Kurdistan is one of the region's most scenic road-trips
- Ahmadawa is a base for treks in the surrounding high country
- The deep Rawanduz Gorge is a popular local day-hike destination

Getting under the skin
Read: International Prize for Arabic Fiction winner *Frankenstein in Iraq* by Ahmed Saadawi; and *The Occupation of Iraq* by Ali A Allawi
Listen: to Kadim Al-Saher, an Iraqi musical megastar
Watch: *Dawn of the World*, directed by Abbas Fahdel; and *My Sweet Pepper Land*, directed by Hiner Saleem
Eat: *masgouf* (barbecued Tigris River fish)
Drink: thick Arabic coffee and sweet black tea

In a word
Salaam aleikum (Peace be with you)

Celebrate this
Iraq is the birthplace of the first great Mesopotamian civilisations, the Sumerians followed by the Akkadian Empire.

Random fact
The Bible's Garden of Eden is believed by some archaeologists to have been in Iraq.

1. Erbil's city centre park and Qaysari Bazaar (covered market) in welcoming Kurdistan
2. The Imam Abbas Shrine in Karbala
3. A baby is baptized in accordance with Yazidi tradition in Kurdistan

Best time to visit

May to September, when the weather is warmer and the days are longer

Top things to see

- The Dingle Peninsula, combining classic craggy Irish coastal scenery with beautiful villages
- Impossibly quaint and photogenic lanes through green fields freshened by bracing ocean winds in Donegal
- The Rock of Cashel's ancient fortifications soaring above Tipperary's plains
- Bobbing boats, narrow lanes and seaside walks around Kinsale
- Dark tunnels and ancient wonders at Neolithic Brú na Bóinne

Top things to do

- Tred the same streets as Ireland's literary greats in Dublin
- Traverse the dramatic surf-sprayed Wild Atlantic Way, linking iconic Irish sights
- Bounce from one venue to the next in rollicking Galway
- Find your own authentic musical moment in a hidden County Clare pub
- Start with the English Market and follow the culinary trail from there in Cork, Ireland's food capital

Getting under the skin

Read: *Angela's Ashes* to understand why Ireland's biggest export for centuries was people; James Joyce to understand the Irish gift for words
Listen: to U2 for sounds bigger than the land; or The Chieftains for sounds of the land
Watch: Roddy Doyle's words come to life in *The Commitments*, a melodic lark; *Brooklyn*, for star turns from Saoirse Ronan and Domhnall Gleeson
Eat: hearty bacon and cabbage; seafood chowder; or smoked salmon and soda bread
Drink: Guinness, possibly chased by a shot of boggy, smoky whiskey

In a word

What's the *craic*? (What's happening?)

Celebrate this

Music is key to understanding Ireland. Punching above its weight in internationally acclaimed musicians (and it's won the Eurovision Song Contest more than any other country), it's the traditional ('trad') scene though that is the foundation of Irish music, inspiring, engaging and moving locals and visitors alike.

Random fact

Ireland is the only country with a musical instrument, the harp, as its national symbol.

1. The Old Library in Dublin's Trinity College; the college counts Oscar Wilde and Samuel Beckett among its alumni

2. Quaint Kenmare in County Kerry in the far southwest of Ireland

3. County Clare meets the cold Atlantic Ocean at the Cliffs of Moher on the west coast

4. Ha'penny Bridge across the Liffey River in Dublin dates from 1816; for just over a century a small toll was charged to cross it

I **CAPITAL** DUBLIN // **POPULATION** 5.2 MILLION // **AREA** 70,273 SQ KM // **OFFICIAL LANGUAGES** ENGLISH & IRISH (GAEILGE)

Ireland

Like a properly poured Guinness (or a simple exchange of pleasantries), this diminutive country flung out in the Atlantic at Europe's far western edge is best enjoyed slowly. That's how to really appreciate the haunting Celtic notes of its traditional music, the lilt to everyday discourse and the intricacies of the countryside – undulating bogs and sheep-flecked green fields criss-crossed by uneven stone walls, silent slate-toned lakes framed by steep-sided valleys, mist-shrouded mountain peaks and filigreed coastline concealing scalloped bays and sandy strands. Even in its lively cities, towns and brightly painted villages, Ireland encourages you to stop, soak up your surroundings and become part of the merriment (or *craic*), beguiled by seemingly nothing at all.

Best time to visit
March to May and September to November

Top things to see
- Jerusalem's walled Old City, dominated by the golden Dome of the Rock
- The West Bank's cliff-side Mar Saba monastery
- Tel Aviv's beach life and Bauhaus architecture
- Caesarea's Roman ruins, one of the great harbours of the classical Mediterranean world
- The Church of the Nativity and the skinny lanes of the old town in Bethlehem

Top things to do
- Trek a section of the Palestinian Heritage Trail through the West Bank
- Float in the Dead Sea then hike the Snake Trail up to Masada
- Cable-car from Jericho to the cliff-hewn Monastery of the Qurantul
- Explore the desert wilderness of the Negev and climb Maktesh Ramon Crater
- Tour the terraced hillside of Baha'i Gardens in Haifa

Getting under the skin
Read: *My Promised Land: The Triumph and Tragedy of Israel* by Ari Shavit; *The 100 Years' War on Palestine* by Rashid Khalidi
Listen: to The Idan Raichel Project
Watch: Oscar-nominated *Omar*, directed by Hany Abu-Assad
Eat: hummus, a national obsession
Drink: beer from the Palestinian Taybeh Brewing Company and award-winning Israeli wines

In a word
Shalom (Hello; peace)

Celebrate this
Jericho vies for title of oldest city in the world, Jerusalem's Mount of Olives is the oldest continuously used cemetery and the church ruins at Megiddo are among the oldest church remains ever excavated.

Random fact
The Dead Sea's water levels are dropping at the rate of about a metre a year, in large part due to water being diverted from the inflowing Jordan River.

1. The gold-plated Dome of the Rock is where, according to Jewish tradition, Abraham prepared to sacrifice his son, while Islamic tradition has the Prophet Muhammad ascending to heaven from this spot

2. Mar Saba monastery is built into the rock face of Kidron Valley in the Judean Desert

3. The Dead Sea is one of the saltiest bodies of water on Earth, almost ten times as saline as the oceans, which is why you will float on it

CAPITAL JERUSALEM (DISPUTED; I); RAMALLAH (P) // **POPULATION** 13.7 MILLION //
AREA 27,660 SQ KM // **OFFICIAL LANGUAGES** HEBREW & ARABIC

Israel & the Palestinian Territories

Israel and the Palestinian Territories are rich with holy sites that hold deep significance for the major monotheistic faiths. The region is also one of the most bitterly contested on Earth, fought over for decades between Israelis and Palestinians. With a backdrop like this, it's almost inevitable that politics ends up colouring many experiences here. But there's more to this place than news headlines. The landscape, diving from rolling, olive-grove hills slashed with deep valleys down to desert only plays second fiddle due to the sheer weight of historic sites. Witnessing scenes of sincere spirituality and discovering the unexpected areas where Arab and Jewish cultures overlap are some of the highlights of a visit.

CAPITAL ROME // POPULATION 62.4 MILLION // AREA 301,340 SQ KM // OFFICIAL LANGUAGE ITALIAN

Italy

As elegant as a finely staged opera, as exuberant as a street carnival, as earthy as a white truffle snouted fresh from the ground – Italy manages to be all things to all people. It's almost a cliche to list Italy's abundant assets, so well known are they on the world stage. Almost any kind of trip you could imagine is possible in this boot-shaped nation, from ski-trips in the icy Alps to cultured city breaks and beach escapes on the shores of Sicily. Whether your tastes run to Roman ruins and Renaissance sculpture or catwalk fashion and Europe's most passionate cuisine, Italy always delivers. The only challenge is how to fit in all the architectural tours, gourmet tastings and languorous, two-hour long lunches.

Best time to visit
Year-round

Top things to see
- Ancient Rome – all of it, but particularly the Colosseum, Forum, Palatine Hill and Pantheon
- Priceless masterpieces in Florence's Uffizi and *David* at the Galleria dell'Accademia
- Baroque Sicily, a treasure-house of masonry flourishes and curling architectural motifs
- Venice's Piazza di San Marco, watched over by the spangled spires of St Mark's Basilica
- The evocative ruins of Pompeii, a thriving commercial town until Mt Vesuvius erupted in 79 CE

Top things to do
- Seek out Italy's finest pizza in the backstreets of Rome or the cobbled lanes of Naples
- Have the drive of your life on the cliff roads of the Amalfi Coast
- Be an adrenaline junkie: ski the Alps, hike the Dolomites, climb the cliffs of Sardinia or mountain bike through the hills around Lake Garda
- Savour a night at the opera at Milan's La Scala
- Make a gourmet pilgrimage through the towns of Piedmont or Emilia-Romagna, sampling legendary cheeses, wines, cold meats and truffles

Getting under the skin
Read: Peter Moore's *Vroom by the Sea: The Sunny Parts of Italy on a Bright Orange Vespa*, a brilliantly written Italy travelogue
Listen: to Andrea Bocelli's uplifting renditions of popular opera classics
Watch: Fellini's classic, *La Dolce Vita*, or *Il Postino*, *Cinema Paradiso* and *Roman Holiday* to get you in the travel mood
Eat: *trippa alla romana* (tripe with potatoes, tomato and pecorino cheese) in Rome; *bistecca alla fiorentina* (T-bone steak) in Florence; and pizza in Naples
Drink: a fine Brunello di Montalcino in Tuscany; Barolo in Piedmont; coffee anywhere

In a word
Ciao! (Hi/Bye!)

Celebrate this
In a packed field of contenders, Leonardo da Vinci stands out as one of Italy's greatest sons: the curiosity and creativity of this Renaissance polymath genius led to innovations in painting, science and more that have had a lasting impact on the world.

Random fact
On average, €3000 a day is tossed into the Fontana di Trevi, Rome's lucky fountain that promises another visit to the capital in exchange for a coin.

1. At the heel of Italy in Puglia, the town of Vieste clings to the Gargano Promontory

2. The Sassolungo mountain range in the Dolomites, seen from Alpe di Siusi cable-car station

3. Bologna in Emilia-Romagna is a great food city but first you need ingredients from Via Pescherie Vecchie in the Quadrilatero

1. The hard-to-navigate Blue Mountains of Jamaica are becoming famous for the quality of the coffee beans produced here

2. East Harbour and Port Antonio lie on the north coast of Jamaica

3. Nine Mile in Saint Ann Parish was Bob Marley's birthplace in 1945

KAROL KOZLOWSKI | AWL IMAGES (ALL IMAGES)

J CAPITAL KINGSTON // POPULATION 2.9 MILLION // AREA 10,991 SQ KM // OFFICIAL LANGUAGE ENGLISH

Jamaica

Jamaica retains strong links to its people's African ancestry. Its most famous flavour, jerk, is a blend of spices evolving from African spice rubs. Its most famous sound, reggae, is an evolution of African folk music. The history of the island based on colonialism and slavery is a brutal one that resulted in a complicated culture, but this Caribbean jewel has rhythm and flavour set against a stunning backdrop where crystalline waters flow over gardens of coral. Red soil and lush banana groves rise into sheer mountains filled with rushing waterfalls. It's a matter of appreciating this great green garden of a land and how its cyclical rhythms set the pace of so much island life.

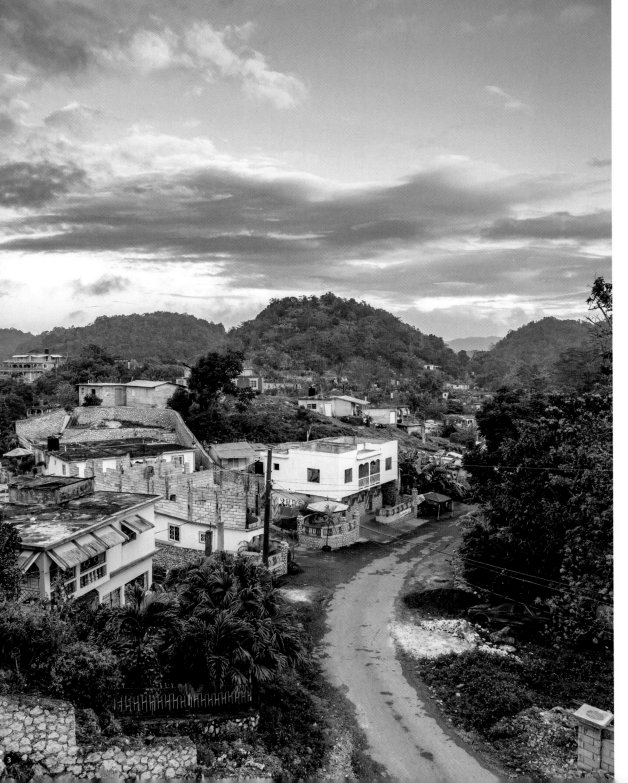

Best time to visit
The weather is beautiful year-round but high season is December to April

Top things to see
- Sunset from the 11km stretch of beach bars and fun at Long Bay on Negril
- The museum dedicated to Bob Marley – see his home and studio and learn about his life
- One of the coffee plantations clinging to cool green mountainsides
- Port Royal, the collapsing former pirate capital of the Caribbean

Top things to do
- Climb 2256m Blue Mountain Peak, part of a lushly forested Unesco World Heritage Site
- Raft the foamy green waters of the Rio Bueno
- Balance on limestone ledges at Dunn's River Falls, which cascade down to a beach
- Get jammin' to the trademark beat of Jamaica: reggae

Getting under the skin
Read: Jean Rhys' *Wide Sargasso Sea*, a tale of post-emancipation Jamaica
Listen: to Bob Marley in his homeland
Watch: 1972's *The Harder They Come*, a cult hit crime flick starring reggae legend Jimmy Cliff, credited with bringing reggae to the masses
Eat: jerk (tongue-searing spice-rubbed barbecued meat)
Drink: the famous Blue Mountain coffee; or the region's greatest variety of rums

In a word
Irie (everything good)

Celebrate this
Kingston is a musical powerhouse churning out some 500 tunes a month and everything from the buses to the dancehalls pulse with the beat.

Random fact
Once the major celebration on the slave calendar, Jonkanoo is a Christmas party in which masked revellers parade through the streets.

J **CAPITAL** TOKYO // **POPULATION** 125.5 MILLION // **AREA** 377,915 KM // **OFFICIAL LANGUAGE** JAPANESE

Japan

Japan is one of those places that both conforms to and confounds expectations. The cliches – Zen gardens, sumo wrestlers, bullet trains, cherry blossom, geisha – are easy to find, but so are experiences that challenge the stereotypes: spotting bears on Hokkaido, basking on tropical beaches in Okinawa Prefecture, skiing on perfect powder in the Japan Alps. Overlaying everything is Japan's complex, nuanced culture: this is a country where manners have been elevated to an art form, and where dressing up as a manga superhero is as commonplace as leaving offerings at a Shinto shrine. Japan manages to be both one of the world's most modern nations, and one of its most traditional – an intoxicating mix for travellers looking for deep immersion into the unfamiliar.

Best time to visit
March to May, to avoid winter snow and summer rain

Top things to see
- Towers, temples and markets piled with tuna in Tokyo, Japan's space-age capital
- Zen gardens, temples, Shinto shrines and geisha in historic Kyoto
- Cherry blossom, repainting Japanese cities in pastel colours every spring
- World Heritage–listed wonders in Nara, the ancient capital of Japan
- Mt Fuji, best viewed from nearby hiking trails or reflected in the Fuji Five Lakes

Top things to do
- Soak in an *onsen* (hot spring) – historic Dogo Onsen in Matsuyama is a top place to plunge
- Be humbled by the lessons of history at Hiroshima and Nagasaki
- Step back to an earlier age in a ryokan (traditional inn) such as Tawaraya in Kyoto
- Strap on skis and spray up some powder on the slopes of the Japan Alps
- See the Japanese casting off inhibitions at spirited festivals such as the semi-nude Hadaka Matsuri

Getting under the skin
Read: Natsume Sōseki's satirical *I Am a Cat*; or Shikibu Murasaki's *The Tale of Genji*, claimed to be the world's oldest novel
Listen: the power pop of Japanese Pop (J-pop) collective AKB48; or the bluesy rock of B'z
Watch: Akira Kurosawa's epic *Seven Samurai*; master animator Hayao Miyazaki's unsettling *Spirited Away*; Hideo Nakata's chilling *Ringu*; or Makoto Nagahisa's indie oddity *We are Little Zombies*
Eat: raw fish, preferably as *sashimi* – slices served with soy, wasabi and preserved daikon radish
Drink: *shochu*, the national spirit of Japan; or *sake*, Japanese rice wine

In a word
Sugoi (The universal exclamation, used whenever something is worth shouting about)

Celebrate this
With Japanese society rapidly ageing, older people are taking the spotlight in all sorts of fields – equestrian Hiroshi Hoketsu, in his seventies, is one of the oldest Olympians ever.

Random fact
Black-skinned Densuke watermelons, often offered as gifts in Japan, can sell for hundreds or even thousands of dollars.

1. Iconic monsters such as Godzilla, here at Hotel Gracery in Shinjuku, Tokyo, are an important part of Japanese culture

2. Kabira Bay, renowned for its white sands and turquoise waters, forms part of the Iriomote-Ishigaki National Park in tropical Okinawa

3. Tokyo's *yokochos* (alleyways) are packed with bars and *izakayas*

4. The original building of Kyoto's Kinkaku-ji dates from 1397 and was a retirement villa for shogun Ashikaga Yoshimitsu

1. The Monastery is one of the key monuments of Petra; similar in design to the Treasury but far bigger, it was built as a Nabataean tomb

2. Wadi Rum National Reserve can be visited on a day trip from Aqaba but camp overnight for the full desert experience

3. Bedouin hospitality: a member of the Beni Atieh tribe welcomes guests with tea

J CAPITAL AMMAN // POPULATION 10.8 MILLION // AREA 89,342 SQ KM // OFFICIAL LANGUAGE ARABIC

Jordan

Adept at manoeuvring in a constantly fluctuating neighbourhood, Jordan is known as the steady hand of the Middle East. The successive huge waves of refugees it has had to absorb – Palestinian, Iraqi and Syrian – makes today's relative stability even more impressive. Something of an ancient world super-highway, Jordan has seen a roll-call of prophets, trading empires and invading armies march through, leaving this country's diverse landscapes, which romp from the deep gash of the Jordan Valley to desert plateau, heaving with historical markers. Petra is the big-hitter site but beyond lay Umayyad desert palaces, Roman ruins, Crusader castles and some of the Holy Land's most important pilgrimage sites.

Best time to visit
March to May or September to November

Top things to see
- The rock-hewn monuments of the Nabataean capital of Petra
- Jerash's Greco-Roman and Byzantine ruins
- The ramparts and vaulted chambers inside the Crusader stronghold of Karak Castle
- Madaba's Byzantine mosaics, and nearby Mt Nebo, where Moses saw the Promised Land
- Amman's Jordan Museum and its 9500-year-old Ain Ghazal statues and the Dead Sea Scrolls

Top things to do
- Camp in Wadi Rum's sand, under a blanket of stars and loomed over by craggy rock outcrops
- Trek from Dana Biosphere Reserve to Petra
- Fail in your attempts to sink in the Dead Sea
- Hike, scramble and swim through the river canyon of Wadi Mujib
- Drive the desert castles loop east of Amman; Qasayr Amra has Umayyad-era frescoes

Getting under the skin
Read: *Pillars of Salt* by Fadia Faqir; and *Land of No Rain* by Amjad Nasser
Listen: to music by Jordanian indie-rock bands Jadal and El Morraba3
Watch: *Theeb,* directed by Naji Abu Nowar; and *Captain Abu Raed,* directed by Amin Matalqa
Eat: *mensaf* (spit-roasted lamb, rice and nuts with a fermented dried-yoghurt sauce)
Drink: *shay* (tea), sometimes served Bedouin-style, flavoured with thyme

In a word
Ahlan wa sahlan (Hello and welcome)

Celebrate this
The Jordan Trail, the first national long-distance trekking route, spans the length of the country.

Random fact
Bethany-Beyond-the-Jordan is where Jesus is believed to have been baptised – it was authenticated by the Pope in 2000.

Best time to visit
May to September

Top things to see
- The glittering blue domes and 15th-century Timurid tilework of the Yasaui Mausoleum in Turkistan
- Kazakhstan's futuristic capital city, Nur-Sultan (formerly Astana), which boasts an indoor tropical beach resort within the world's largest tent
- Cosmopolitan Almaty, with Orthodox cathedrals and leafy sidewalk cafes
- Fishing boats marooned in the desert seabed of Aralsk, miles from the nearest waters of the Aral Sea

Top things to do
- Take a horse trek to the base of Mt Belukha in the magnificent Altay Mountains
- Hike to the three Kolsay Lakes in the southeastern Zailiysky Alatau range
- Glide around Medeu's world-class ice rink or snowboard at nearby Shymbulak ski resort
- Spot flamingos in their most northerly habitat at Korgalzhyn Nature Reserve
- Buy some birch twigs and give yourself a good lashing at Almaty's Arasan Baths

Getting under the skin
Read: *Dark Shadows* by Joanna Lillis, an excellent and up-to-date journalistic portrait of independent Kazakhstan
Listen: to *Aralkum* by Galya Bisengalieva, the Kazakh-born violinist and composer's mournful classical elegy to the dying Aral Sea
Watch: Kazakh hordes battle the Dzungarian armies in Sergei Bodrov's *Nomad*, Kazakhstan's US$40-million blockbuster
Eat: *kazy* (smoked horsemeat sausage) – the ultimate nomad snack
Drink: *shubat* (fermented camel's milk); *kymyz* (fermented mare's milk); or play it safe with a local Derbes beer

In a word
Salametsyz be? (Hello/How are you?)

Celebrate this
In 2020, three of Kazakhstan's estimated 150 snow leopards were filmed strolling just outside Almaty's city limits, near the popular hiking trails of the Bolshoe Almatinskoe Lake.

Random fact
Kazakhstan sits atop an estimated 100 billion barrels of oil, most of it along the shores of the Caspian Sea.

1. Bayterek Monument in Nur-Sultan embodies a Kazakh legend in which the mythical bird Samruk lays a golden egg containing the secrets of human happiness in a tall poplar tree

2. Yasaui Mausoleum in Turkistan in the south of the country commemorates a Turkic poet and mystic

3. The mountainous landscape of the Kok-Zhailau plateau near Almaty

K CAPITAL NUR-SULTAN // POPULATION 19.1 MILLION // AREA 2,724,900 SQ KM // OFFICIAL LANGUAGES KAZAKH & RUSSIAN

Kazakhstan

Gargantuan Kazakhstan is the world's ninth-largest country, but it remains a mystery to many travellers. Its big draws are definitely the superb Altay and Tian Shan mountains bordering China, but the bleak, bewildering steppe also beckons with surreal, surprising secrets ranging from Soviet-era cosmodromes to underground mosques and the rusting ruins of the Aral Sea. After years of collectivisation, the former horsemen of the Golden Horde are now getting rich on petrodollars, while trying to deal with the legacy of serving as the Soviet Union's favourite dumping ground. Today the Kazakh steppe offers one of the last great undiscovered frontiers of travel.

CAPITAL NAIROBI // **POPULATION** 53.5 MILLION // **AREA** 580,367 SQ KM // **OFFICIAL LANGUAGES** KISWAHILI & ENGLISH

Kenya

The earth moved when Kenya was created. And the evidence is thankfully clear for all to see, with the Great Rift Valley cutting dramatically across the country's length – its escarpments, lakes, savannahs and volcanoes are some of Africa's most beautiful landforms. They also form habitats for an astounding diversity of iconic African wildlife. Safari ecotourism and community conservancies in the Masai Mara and Laikipia are protecting more ecosystems than ever before, as well as helping to ease modern pressures on the traditional cultures of peoples such as the Maasai, Samburu, Turkana, Kikuyu and Luhya. The Swahili coast's beaches and the island of Lamu offer another take on Kenya, as does Nairobi – the business and technological heart of East Africa's powerhouse modern economy.

Best time to visit
January to February, June to October

Top things to see
- The 'Great Migration' crossing the Mara River, when hundreds of thousands of wildebeest and zebras dodge the waiting jaws of crocodiles
- Lake Turkana (aka Jade Sea) shimmering in the desert of northern Kenya
- Black rhinos and rare Grevy's zebras in the outstanding Lewa Wildlife Conservancy
- Some of East Africa's largest elephant herds in Amboseli National Park, with Kilimanjaro as a backdrop
- Swahili culture in all its glory within the enigmatic environments of Lamu

Top things to do
- Spend five epic days trekking up to Point Lenana (4979m) on Mt Kenya for sunrise
- Rent a bicycle and pedal through the wilds and wildlife of Hell's Gate National Park
- Hike along the serrated crater rim of Mt Longonot and look down into a forgotten world
- Stroll through clouds of butterflies and spot primates and countless birds in Kakamega Forest
- Learn how to harness the winds during kitesurfing lessons on Diani Beach, south of the historic city of Mombasa

Getting under the skin
Read: *A Grain of Wheat* by Ngũgĩ wa Thiong'o, which is set at the time of Kenya's independence; *Unbowed: A Memoir* by 2004 Nobel Peace Prize laureate Wangari Maatha
Listen: to *Kilio Cha Haki - Ukoo Fulani Mau Mau*, a ground-breaking album featuring 38 Kenyan artists
Watch: *Kati Kati*, a story of afterlife in the Kenyan wilderness; *Rafiki*, a story of love between two young women (the first Kenyan film screened at Cannes Film Festival)
Eat: *nyama choma* (Kenyan-style roasted meat), a perfect accompaniment to cold beer
Drink: milk mixed with cow's blood at a Maasai celebration – if you dare!

In a word
Hakuna matata (No worries)

Celebrate this
Eliud Kipchoge, already the marathon world record holder (2:01:39), did the unthinkable on 12 October 2019 – breaking the two-hour barrier, with a 1:59:40 marathon in Vienna, Austria.

Random fact
Nairobi National Park is the only national park in the world to be in a capital city – on its savannah plains it is possible to view rhinos, lions, giraffes and other safari species with the city's skyline as a backdrop.

1. Beadwork worn by women is an important part of Maasai culture and each colour holds a meaning

2. Elephants amble in the shadow of Mt Kilimanjaro in Amboseli National Park

3. To understand Kenyan culture, visit the National Museum in Nairobi, East Africa's greatest city

1. Find corals and tropical fish below the surface of the Pacific around Tabuaeran atoll, although over-fishing has damaged the biodiversity

2. One of the many low-lying atolls of Kiribati, showing their vulnerabilty to rising sea levels

3. Floating on flotsam after a high tide flood in the village of Eita

Kiribati

Kiribati (pronounced 'Kiribas'), with its aqua lagoons and sensational sunsets, is equally as alluring for its real-world simplicity. Isolated by the vast Pacific Ocean and untainted by tourism, the country is made up of 33 low-lying islands and atolls flung across the equator, putting it on the front line of climate change. The sound and sight of the sea dominates here, fish are the staple and boats are the major form of transport. Locals might wonder why you've come to visit, but with a friendly attitude you'll be welcomed with warm smiles. Explore densely populated and increasingly modern Tarawa, or get completely back to Kiribati roots on the outer islands. Wherever you go, don't expect schedules or luxury; just go with the flow.

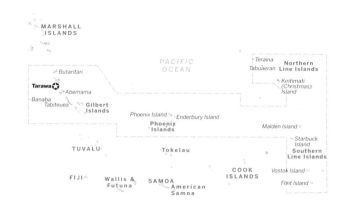

ZEYNEPIZTURK · GETTY IMAGES // GAIL SHOTLANDER | GETTY IMAGES //
JONAS GRATZER | GETTY IMAGES

Best time to visit
March to October, the dry season, when it's marginally less like living in a sauna

Top things to see
- Traditional dances that emulate the movements of the frigate bird
- An abundance of seabirds, crabs and turtles at the isolated Phoenix Islands Protected Area
- Rusted tanks, ships and aeroplanes on the reef, all remnants from the WWII Battle of Tarawa
- Kiribati crafts from the outer islands including *te wii ni bakoa* (hand-smocked tops) and conical woven pandanus fishermen's hats

Top things to do
- Saltwater fly-fish for famously feisty bonefish
- Surf lonely reef breaks English Harbour and Whaler's Anchorage off Tabuaeran (Fanning Island)
- Snorkel the sublime waters off Christmas Island
- Salt or smoke your catch and learn self-sufficiency in the outer islands

Getting under the skin
Read: *Kiribati* by Alice Piciocchi, which captures Kiribati culture amid the climate change emergency
Listen: to singing emanating from *maneaba* (traditional meeting houses) across the country
Watch: *Tinau* (My Mother), a documentary on the threat of rising sea levels
Eat: fresh fish, breadfruit and rice
Drink: *kaokioki*, also called sour toddy, made from fermented coconut palm sap

In a word
Mauri-i-Matang! (Hello stranger!)

Celebrate this
Kiribati's pristine Phoenix Islands Protected Area, equal to the size of California, is the world's largest designated Marine Protected Area. It's home to 14 extinct underwater volcanoes and hundreds of coral, fish and marine mammal species.

Random fact
Kiribati is spread over 3.55 million sq km of ocean, the largest ocean-to-land ratio in the world.

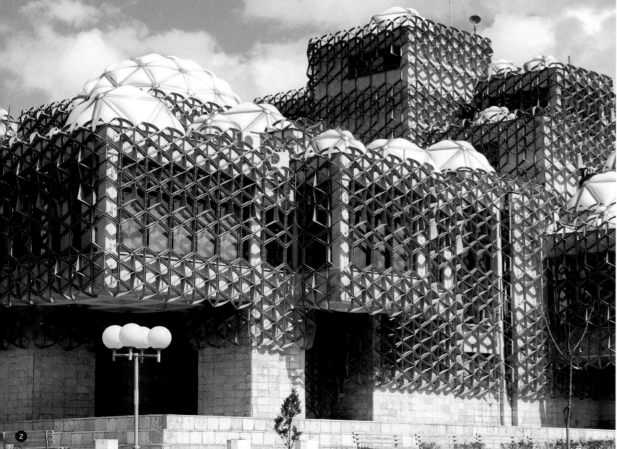

Best time to visit
May to September

Top things to see
- Old Prizren, with its hilltop castle, arched Ottoman bridge, Sinan Pasha Mosque and Mehmed Pasha Baths
- The serenity of Dečani Monastery, whose interior is like an enormous medieval paintbox
- Extraordinary stalagmites in Gadime Cave, formed over millions of years
- Slivers of sunlight carving through the atmospheric gloom in Gračanica Monastery

Top things to do
- Talk up a storm with the young and expat crowd in the bars and cafes of Pristina
- Visit the rescued brown bears at the Bear Sanctuary Pristina, where formerly caged bears are kept in spacious, semiwooded enclosures
- Test your nerves at the challenging Via Ferrata Berim on Mokra Gora mountain
- Pull on your walking shoes in the wild mountain scenery of Rugova Valley and the Shar Mountains
- Hit the slopes skiing at Brezovica, Kosovo's premier destination for winter sports

Getting under the skin
Read: Albanian writer Ismail Kadare's *Three Elegies for Kosovo*, a tale of the land's sad history
Listen: to traditional folk music featuring skirling flutes and goat-skin drums
Watch: Isa Qosja's *Three Windows and a Hanging*, addressing the topic of raped women in a patriarchal society
Eat: a variety of Balkan staples including Turkish baklava, Serbian *ćevapčići* (grilled kebab) or *kos* (goat's-cheese yogurt)
Drink: tea by the glassful at a traditional teahouse; a fine selection of macchiatos in cafe-packed Pristina; or locally brewed Peja beer almost anywhere

In a word
Tungjatjeta/Zdravo (Hello in Albanian/Serbian)

Celebrate this
Kosovo alpinist Uta Ibrahimi was the first Albanian woman to climb Mt Everest, and the first woman from the Balkans to summit five 8000m peaks. She is UN Kosovo's Sustainable Goals Champion, promoting gender equality, environmental protection and youth empowerment.

Random fact
International pop stars Rita Ora and Dua Lipa both have ethnic Albanian roots and parents who moved from Kosovo to the UK.

1. The glacial lake of Liqeni i Drelajve in the Prokletije mountain range on the Kosovo-Montenegro border

2. Kosovo's National Library was built in 1982 by the Croatian architect Andrija Mutnjakovic

3. Prizren is one of Kosovo's oldest settlements seeing both Roman and Ottoman rule

Kosovo

Synonymous still with the brutal ethnic war that happened here in the late '90s, Kosovo was one of the last pieces of the former-Yugoslav puzzle to secede, and is now Europe's newest country, having declared independence in 2008. While Kosovo's status remains contested by many, most particularly by next-door Serbia, for whom the former province is considered holy ground, it's a functional and safe place to visit these days, with EU and UN agencies still present. For travellers Kosovo offers superb mountain scenery and hiking, the charming old quarters of Prizren and Peja (Peć), several ancient NATO-protected Orthodox monasteries and a fast-changing (if rather slapdash) capital in Pristina.

1. Kuwait City's stripy water towers began work in 1977, storing essential water from a desalination plant

2. Kuwait's Grand Mosque can accommodate 10,000 men in its main prayer hall and 950 women in a separate hall

3. Kuwait's tallest building, the Al Hamra tower, was completed in 2011

CAPITAL KUWAIT CITY // POPULATION 2.9 MILLION // AREA 17,818 SQ KM // OFFICIAL LANGUAGE ARABIC

Kuwait

A tiny city state hemmed in by two Middle Eastern giants, Kuwait occupies one of recent history's most contested regions. That is as true today as it was yesterday. But civilisation here dates back millennia to the Bronze Age Dilmun civilisation, one of the world's oldest trading commnities. Fast forward to the present and Kuwait wears its oil riches lightly – though developments such as 2019's opening of the 48.5km Sheikh Jaber Al-Ahmad Al-Sabah Causeway across Kuwait Bay demonstrate that it doesn't shy away from ambitious projects. Both a bastion of Gulf tradition and an emerging battleground for women's rights and liberalising trends, Kuwait is the Gulf's past, present and future in one small space.

Best time to visit
November to April

Top things to see
- The extraordinary collection of ethnographic artefacts, textiles and Arabic manuscripts at Tareq Rajab Museum
- Kuwait Grand Mosque's interior, a contemporary interpretation of classical mosque design
- Dar Al-Athar Al-Islamiyya, to view the Al-Sabah collection of Islamic art
- The Mirror House, to tour the kooky mirror-mosaic-covered creation of artist Lidia Al-Qattan

Top things to do
- Zoom up to the viewing deck of the Kuwait Towers for city vistas
- Bargain for kitsch and search for treasure in sprawling Souq Mubarakiya
- Head to the Mutla Ridge for quad biking and desert-plateau views

Getting under the skin
Read: *The Bamboo Stalk* by Saud Alsanousi; and *The Pact We Made* by Layla AlAmmar
Listen: to Abdullah al-Rowaishid, who blends traditional music with Arabic pop
Watch: *Fires of Kuwait* by David Douglas, which follows the teams cleaning up burning oil wells after the Iraq war
Eat: Gulf fish baked with coriander and cardamom
Drink: *lumi* (dried black-lime) tea

In a word
Marhaba (Hello)

Celebrate this
Although nearly the entire Al-Sabah collection of Islamic art was stolen by withdrawing Iraqi forces in 1991, most was returned after UN pressure. Highlights are displayed in the Dar Al-Athar Al-Islamiyya.

Random fact
During the Iraqi occupation, Tareq Rajab Museum's custodians bricked up the doorway and littered the entrance steps with rubbish, fooling would-be looters into going no further.

K CAPITAL BISHKEK // POPULATION 6.0 MILLION // AREA 199,951 SQ KM // OFFICIAL LANGUAGES KYRGYZ & RUSSIAN

Kyrgyzstan

A land of mountain valleys, glittering lakes and felt yurts, Kyrgyzstan is a dream for DIY adventurers, responsible tourists and aspiring nomads (visit immediately if you are all three). After the collapse of the USSR, tiny Kyrgyzstan turned to tourism, creating a cutting-edge network of community-based ecotourism ventures and homestays. Dozens of adventures await the intrepid, from horse treks and yurt stays to visits to eagle-hunters, with your tourist dollars going direct to local families. Throw in some Silk Road bazaars and a tradition of nomad-inspired hospitality, and most who have visited agree that Kyrgyzstan is Central Asia's most enjoyable destination.

Best time to visit
June to September

Top things to see
• Issyk-Köl, a huge inland sea fringed with beaches and framed by snowy peaks
• Tash Rabat, Central Asia's most evocative caravanserai (historic inn for Silk Road merchants)
• Kyrgyzstan's second city of Osh; an ancient Silk Road bazaar town on the edge of the Fergana Valley
• The national sport of *ulak tartysh*, like rugby on horseback, and known affectionately among travellers as 'goat polo'
• The production process for *shyrdak* (felt carpets), Kyrgyzstan's most iconic souvenir

Top things to do
• Live like a nomad on a horse trek to remote Song-Köl lake
• Overnight in a yurt or community-tourism homestay to gain an insight into traditional local life
• Trek through the pristine alpine valleys of the Tian Shan range near the city of Karakol
• Go heli-skiing in the Tian Shan mountains, just an hour from the capital, Bishkek
• Cross into China over the Irkeshtam or Torugart passes, Central Asia's most exciting border crossings

Getting under the skin
Read: the Kyrgyz novel *Jamilla* by Chingiz Aitmatov, Central Asia's best known novelist
Listen: to *Music of Central Asia Vol 1: Mountain Music of Kyrgyzstan*, (Smithsonian Folkways), a playlist of Kyrgyz music from traditional ensemble Tengir-Too
Watch: Bekzat Pirmatov's 2018 *Aurora*, set in the sanatorium of the same name at Issyk-Köl
Eat: *beshbarmak* ('five fingers'), a traditional dish of flat noodles and mutton, cooked in broth and eaten by hand
Drink: *kumys* (fermented mare's milk), sold along country roads in spring and summer

In a word
Ishter kanday? (How are things?)

Celebrate this
In 2019 the distinctive Kyrgyz white felt hat known as an *ak kalpak* was added to Unesco's list of intangible cultural heritage, to the immense pride of Kyrgyz men everywhere.

Random fact
The Kyrgyz oral poem, the *Epic of Manas*, is the world's longest poem and is 20 times longer than the *Odyssey*.

1. Explore the valleys around Karakol in the Tian Shan mountain range, which spans several countries

2. Catch a game of *ulak tartysh* as mounted players try to place a goat carcass in the opposing team's goal

3. Selling traditional and not-so-traditional rugs at a nomadic village on the Silk Road

L CAPITAL VIENTIANE // POPULATION 7.4 MILLION // AREA 236,800 SQ KM // OFFICIAL LANGUAGE LAO

Laos

Thanks to its laid-back way of life and impressive hand of historic sights – the temples of Luang Prabang, French-influenced Vientiane, the sprawling ruins of Wat Phu, the Plain of Jars at Xiengkhouang (recently added to the World Heritage list) – this fascinating Buddhist nation is at the top of many a backpacker's bucket list. Today, Laos is growing in popularity as the new backdoor to Southeast Asia. Perhaps appropriately, escaping the modern age is the most popular thing to do, whether that means trekking in remote regions, enjoying punchy, authentic Laotian cuisine, taking to the trees to spot gibbons near Huay Xai, or observing the surging waters of the Mekong from a hammock at Si Phan Don.

Best time to visit
November to February, to avoid the worst of the humidity

Top things to see
- The distinctive, angular stupa of gold-covered Pha That Luang in Vientiane
- One monk's vision of heaven and hell in the sculpture garden at Xieng Khuan
- Irrawaddy dolphins splashing around 4000 river islands at Si Phan Don
- A garden of jewelled monasteries in World Heritage–listed Luang Prabang
- Rafting, caving and climbing around dramatic karst outcrops at Vang Vieng

Top things to do
- Drop into a Lao *wat* (Buddhist temple-monastery) for a chat with the novices
- Take the slow boat along the Mekong from Luang Prabang to Nong Khiaw
- Stay in a homestay on a trek through Nam Ha National Protected Area
- Explore caverns once used by Pathet Lao guerrillas at Vieng Xai
- Cross the forest canopy by zip line while tracking black gibbons near Huay Xai

Getting under the skin
Read: Brett Dakin's *Another Quiet American* or Dervla Murphy's *One Foot in Laos* for a personal take on the PDR (People's Democratic Republic)
Listen: to the undulating melodies of the *khene* (traditional reed pipe), or the urgent rock of CELLs
Watch: *Good Morning, Luang Prabang*, the first ever privately funded Lao movie; or eerie fantasy *The Long Walk* directed by Mattie Do
Eat: *laap* (spicy, marinated meat salad); or *tam maak hung* (green papaya salad)
Drink: *lao-lao* (rice liquor); or Beer Lao, the nation's favourite brew

In a word
Su kwan (The calling of the soul)

Celebrate this
Laos is one of the few countries where homosexuality has never been criminalised. Since the first official gay pride march in 2012, gay lives have become increasingly visible, thanks to the work of LGBTI organisations such Proud to Be Us Laos.

Random fact
The Lao people are the world's most prolific consumers of sticky rice, getting through 155kg per person per year.

1. Sunset over an Akha hilltribe village in northern Laos near Luang Namtha

2. The golden stupa of Pha That Luang in Vientiane is believed to enshrine a breast bone of the Buddha

3. Southeast of Vientiane, eccentric Xieng Khuan features other-worldly Buddhist and Hindu sculptures

4. Wat Xieng Thong in Luang Prabang has a a stunning 'tree of life' mosaic on its west wall

Best time to visit
May to September

Top things to see
- The medieval castle complex in Cēsis and its elegant 18th-century manor
- Enchantingly desolate and beautiful Cape Kolka
- Rūndale Palace, Latvia's miniature version of Versailles (but without the crowds)
- Hilltop Turaida Museum Reserve with its medieval ramparts and verdant walking trails
- Liepāja's windswept shores and art nouveau buildings
- Rīga's Museum of Occupation and medieval core

Top things to do
- Uncover emerald lakes and blueberry fields in the Latgale Lakelands
- Bobsled down a 16-bend track at 80km/h in high-adrenaline Sigulda
- Hobnob with Russian jetsetters in the heart of Jūrmala's swanky spa scene
- Snack your way around the zeppelin hangars of Rīga's bounteous Central Market
- Stumble upon yawning caves, nature trails and Soviet bunkers in Gauja National Park

Getting under the skin
Read: *The Merry Baker of Riga* by Boris Zemtzov, an amusing tale of an American entrepreneur setting up shop as a baker in Rīga in the early 1990s
Listen: to Prāta Vētra (aka Brainstorm) for pop/rock; synth-heavy rock from the reborn Otra Puse
Watch: Jānis Streičs' *The Child of Man*, about a boy growing up and falling in love in Soviet-occupied Latvia; and *The Mystery of the Old Parish Church*, tackling the prickly issue of locals collaborating with Nazi and Soviet occupiers during WWII
Eat: the almighty pig and ubiquitous potato; sausages; smoked fish; soups of beets, nettles and sorrel with dark rye bread; freshly picked berries in summer, mushrooms in autumn
Drink: Latvia's Black Balzām, a jet-black, 45% proof concoction that Goethe called 'the elixir of life'

In a word
Sveiki (Hello)

Celebrate this
Latvia, especially Rīga and Liepāja, fell in love with art nouveau in the early 20th century; some 30% of central Rīga is built in the style, mostly apartment blocks, giving residents and visitors a visual treat literally round every corner.

Random fact
Held every five years, the Latvian Song and Dance Festival unites close to 40,000 participants in a jaw-dropping, sweet-sounding display of patriotism.

1. Rundāle Palace in Pilsrundāle was designed by Italian Bartolomeo Rastrelli, who is also known for the Winter Palace in St Petersburg

2. Enjoy an evening organ recital in Rīga Cathedral

3. Refuel at Lielezers, a traditional Latvian bakery stall at Rīga's Central Market

4. Gauja River on an icy winter day at Sigulda

L **CAPITAL** RIGA // **POPULATION** 1.9 MILLION // **AREA** 64,589 SQ KM // **OFFICIAL LANGUAGE** LATVIAN

Latvia

If you've an appetite for Europe's lesser-known destinations, a taste of Latvian life should stimulate the senses. Tucked between Estonia to the north and Lithuania to the south, Latvia is the meat of the Baltic sandwich, the savoury middle, loaded with colourful fillings. Thick greens take the form of Gauja Valley pine forests peppered with castle ruins – though these days as many people visit for adventure sports as medieval treasures. Onion-domed orthodox cathedrals cross the land from salty Liepāja to gritty Daugavpils. Cheesy Russian pop blares along the beach in Jūrmala. And capital Rīga adds an extra zing as the country's cosmopolitan nexus, serving some sweet eye candy: Europe's largest and loveliest collection of art nouveau architecture, and cobbled lanes hidden behind gingerbread trim.

Best time to visit
March to November

Top things to see
- Baalbek's collection of temples, extravagantly decorated and built on a giant's scale
- Beirut's museums and art galleries and the offshore limestone arches of Pigeon Rocks
- Byblos, an ancient harbour town with a vast archaeological site and Crusader castle
- The Qadisha Valley's Unesco World Heritage–listed rock-hewn Maronite monasteries
- Tyre's ruins, vestiges of one of the ancient Mediterranean's most important port cities

Top things to do
- Explore the 6km stretch of colossal stalactites and stalagmites inside Jeita Grotto
- Ski or tandem-paraglide in The Cedars
- Hike through the Chouf Mountains, arguably Lebanon's most spectacular scenery
- Get lost amid the maze of lanes in Saida's souq

Getting under the skin
Read: *Lebanon: A Country in Fragments* by Andrew Arsan; and Rabee Jaber's novel *The Mehlis Report*
Listen: to Fairuz, icon of Middle Eastern music; and Mashrou' Leila, Lebanon's biggest rock band

Watch: Oscar-nominated *Capernaum,* directed by Nadine Labaki; and *West Beirut*, directed by Ziad Duweyri
Eat: Lebanese *mezze* (small plates) spreads including *kibbeh* (lamb croquettes), *muttabal* (eggplant and tahini dip) and *warak anaib* (stuffed vine leaves)
Drink: Lebanese wines; *arak* (aniseed-flavoured spirit); Almaza (the local beer)

In a word
Ahlan wa sahlan (Hello and welcome)

Celebrate this
Lebanon's cedars are the world's oldest historically documented forests. Although few of the once sprawling woods remain, some of the surviving trees are estimated to be around 2000 years old.

Random fact
The southern Lebanese coast around Tyre and Sidon is where the ancient Phoenician Empire was born.

1. Known as the Heliopolis or 'Sun City' of the ancient world, Baalbek's ruins, such as the Bacchus Temple, are some of the best preserved in the Middle East

2. Catching a movie at a new drive-in cinema in Byblos during the coronavirus pandemic

3. Sursock Palace, dating from the Ottoman era, was damaged during the huge explosion in Beirut in 2020

4. Bsharri is a town in the historic Qadisha Valley and also home to Lebanon's oldest ski resort

CAPITAL BEIRUT // **POPULATION** 5.4 MILLION // **AREA** 10,400 SQ KM // **OFFICIAL LANGUAGE** ARABIC

Lebanon

Lebanon's landscapes – swooping from snow-capped mountains, through valleys and down to the Mediterranean coast – are as diverse as its population, which officially incorporates 18 religious sects while also absorbing 1.7 million Syrian and Palestinian refugees. Often feted as the Middle East's most urbane corner, Beirut is known for its nightlife, dining and cafe scene. Outside of the capital though, ancient ruins and fortress remnants remain and there are stone-cut villages and wineries huddled in the hills and harbour-town souqs to explore. Diversity is Lebanon's strength but also a vulnerability. While often used as a football by regional neighbours, it's Lebanon's civil war hangover of sectarian grievances that has fuelled corruption executed by its political fiefdoms. This was made ever more apparent by Lebanon's now long-running economic crisis and 2020's deadly Beirut port explosion.

Best time to visit
May to September, to avoid the rains and mist

Top things to see
- Fossilised dinosaur footprints while hiking around Quthing, Leribe and Morija
- Water plummeting 204m from the top of Maletsunyane Falls
- San rock paintings around Malealea and the aptly named Gates of Paradise Pass
- Thaba-Bosiu (Mountain at Night) – the old stronghold of King Moshoeshoe the Great and the birthplace of the Basotho nation
- Lovely mountain panoramas around the Moteng Pass and Oxbow

Top things to do
- Saddle up and ride a Basotho pony through Lesotho's rugged interior
- Experience the silence, isolation and rolling grasslands of Sehlabathebe National Park
- Ascend to the 3090m-high Mafika-Lisiu Pass for a never-ending vista
- Sleep in a traditional trading post for a glimpse of authentic village life
- Hike in wild and beautiful Ts'ehlanyane National Park or Bokong Nature Reserve

Getting under the skin
Read: *Basali! Stories by and about Women in Lesotho,* edited by K Limakatso Kendall; *Everything Lost Is Found Again: Four Seasons in Lesotho*, by Will McGrath
Listen: to the *lekolulo*, a flute-like instrument played by herd boys
Watch: *This Is Not a Burial, It's a Resurrection*, the story of a widow defending her village's spiritual heritage; *The Forgotten Kingdom*, which follows a young man returning to Lesotho to bury his father
Eat: *papa* (maize meal) and *moroho* (greens)
Drink: *joala* (traditional sorghum beer) – a white flag flying in a village means that it's available

In a word
Khotso (Peace)

Celebrate this
Mathabiso Mosala: she was at the forefront of Lesotho National Council of Women for over half a century, and worked to provide women with meaningful skills, opportunities and training; fittingly, her granddaughter is Director-General of the Lesotho National Broadcasting Service (LNBS), who in turn is putting great effort into creating equal opportunities for people from disadvantaged backgrounds.

Random fact
Lesotho is the only country on earth that exists entirely above 1000m; its lowest point is a lofty 1400m (4593ft).

1. Cattle are herded past village rondavels

2. In Basotho culture, the blanket is an indispensable item of clothing

3. The Maletsunyane Falls are almost twice the height of Victoria Falls

L CAPITAL MASERU // POPULATION 2.0 MILLION // AREA 30,355 SQ KM // OFFICIAL LANGUAGES SESOTHO & ENGLISH

Lesotho

Set in the sky and surrounded by South Africa, the mountainous Kingdom of Lesotho is the only stronghold for the Basotho people. Although its vastly larger neighbour is just a mountain pass away, the high peaks and isolated valleys of the 'Kingdom in the Sky', and the pride of the Basotho in their identity, have served to insulate Lesotho's culture. Outside of Maseru's thin urban veneer, life quickly gives way to traditional customs and nature. Herd-boys tend their sheep on steep hillsides, horsemen wrapped in *kobo* (Basotho blankets) ride over high mountain passes and village festivals are a focal point of local life. Witnessing Lesotho, whether trekking or on the back of a famed Basotho pony, is a worthwhile adventure.

Ⓛ **CAPITAL** MONROVIA // **POPULATION** 5.1 MILLION // **AREA** 111,369 SQ KM // **OFFICIAL LANGUAGE** ENGLISH

Liberia

Founded by a few thousand freed enslaved African-Americans in the early 19th century, but inhabited for far longer by traditional groups famous for their artistic traditions and secret societies, Liberia's complicated cultural mix hasn't always worked. Without a right to vote for well over a century, and enduring forced labour for almost as long, original inhabitants (who've always made up at least 95% of the population) did not see a president that wasn't of African-American ancestry until 1980. Now, almost two decades since its last civil war and led by a new generation of politicians, Liberia is at peace and hope abounds. Its people and its natural spectacles – splendid beaches, barely penetrable rainforest – are all calling for you.

Best time to visit
November to April (the dry season)

Top things to see
- The village-turned-city, Monrovia, with the ruins of war and the frenetic activity of peace
- Forest elephants, pygmy hippos and chimps within the rainforest of Sapo National Park
- Harper, with its charming southern American architecture and end-of-the-line feel
- Firestone Plantation, where you can learn about rubber-tapping, one of Liberia's original industries

Top things to do
- Share the perfect Atlantic breaks with local surfers off Robertsport
- Climb up Mt Nimba (1362m), Liberia's tallest peak, which is baggable in a few days from Monrovia
- Camping on the wild beaches of Buchanan, the vibrant yet tranquil port city
- Picnic at the pretty semicrater lake of Bomi, near Tubmanburg

Getting under the skin
Read: *She Would Be King*, a reimagining of Liberia's early history; *Madame President: The Extraordinary Journey of Ellen Johnson Sirleaf*, by Helene Cooper
Listen: to *We Want Peace* by Gebah and Maudeline (The Swa-Ray Band); Cralorboi CIC, for his engaging Afro-pop
Watch: *Johnny Mad Dog*, which tells the story of child soldiers in Liberia's second civil war; *Freetown*, based on a true story of missionaries trying to escape Liberia's first civil war
Eat: goat soup and traditional rice bread made with mashed bananas
Drink: ginger beer; *poyo* (palm wine); and strong coffee

In a word
Peace, man

Celebrate this
National Unification Day (14 May), which aims to tackle political, social and economic disparities within the country.

Random fact
Liberia is the only nation in the world to have had presidents that have both won the Nobel Peace Prize (Ellen Johnson Sirleaf) and the Ballon d'Or (George Weah).

1. Liberians celebrate on Monrovia's beach as the country's Ebola epidemic subsides

2. A game of football in West Point township; the median age in Liberia is just 19 years

3. Heading towards the mountains and Mt Nimba via Ganta

Best time
October to March, for cooler temperatures

Places of interest
- The Roman city of Leptis Magna strung out along the Mediterranean
- Cyrene, where Ancient Greece and Rome have been grafted onto African soil
- Ghadames, the Sahara's most captivating oasis and caravan town
- Cosmopolitan capital city, Tripoli, with an Ottoman-era medina and world-class museum
- Jebel Acacus, the haunting desert massif in the Sahara's heart with 12,000-year-old rock art

Local customs
- Dipping three fingers into a bowl of perfumed water prior to meals as a form of ritual cleansing
- Libyans love packing a picnic and heading for a beach along the Mediterranean coastline
- Meals are of great symbolic importance, with lunch being the most significant
- Although having adopted Islam, Tuareg men – instead of women – cover their faces in public

Getting under the skin
Read: *In the Country of Men* by Hisham Matar, a searing novel about modern Libya; *The Burning Shores: Inside the Battle for the New Libya*, by Frederic Wehrey; *Gold Dust* by Ibrahim al-Koni

Listen: to *malouf*, a traditional musical form from Andalucía that's the accompaniment to most celebrations; Benghazi-born Hamid El Shaeri, for his Al-jil (synthesizer pop)

Watch: *Hondros*, a documentary about photographer Chris Hondros who was killed covering the civil war in 2011; *The English Patient*: it may not have been filmed in Libya, but this Oscar-winning epic captures the essence of old Libya from Tobruk to the Sahara

Eat: couscous, the national dish; home-made bread, cooked beneath the sand and enjoyed under Saharan stars

Drink: three glasses of strong, sweet tea with the Tuareg around a campfire in the Sahara

In a word
Bari kelorfik (Thank you) – a blessing

Celebrate this
Hajer Sharief, a human rights activist and law student, was nominated for the Nobel Peace Prize in 2019 for her work pushing for peace and equal representation of women and young people in politics.

Random fact
For 90 years Libya was famed for holding the world record for the highest air temperature recorded (57.8°C), until the claim was rejected by the World Meteorological Organisation in 2012.

1. The port of Leptis Magna was one of the Roman Empire's greatest cities, until conquered by a Muslim army in 647

2. A coppersmith works in Tripoli's Ottoman-era medina

3. When Italian forces returned to Libya in the 20th century they set about rediscovering the ruins of Leptis Magna

4. The Berber oasis of Ubari Lakes with an ocean of Saharan sand dunes beyond

L   **CAPITAL** TRIPOLI // **POPULATION** 6.9 MILLION // **AREA** 1,759,540 SQ KM // **OFFICIAL LANGUAGE** ARABIC

Libya

An ancient crossroads of great civilisations, Libya today finds itself and its people at its own crossroads and in the crosshairs. The post-Gaddafi era continues to bring violent instability, with armed militias having divided the country into a series of mini-states. It will undoubtedly be some years before Libya is safe to visit. When that changes, the nation will once again welcome travellers with a stunning coastline, relics of Roman and Greek cities, poignant WWII sites, remote massifs adorned with prehistoric rock art, and some of the most beautiful stretches of the Sahara Desert, where the shifting sands serve as a reminder that nothing stays the same forever.

Best time to visit
December to March for winter sports; May to late September for summer hikes

Top things to see
- Schloss Vaduz, the turreted fairy-tale castle looming over the capital
- Postage stamps issued in the principality since 1912 at the Post Museum in Vaduz
- The snowy slopes of Malbun where British royal Prince Charles learnt to ski
- Austria, Switzerland and most of little Liechtenstein from the lofty crests of Malbun's circular Fürstin-Gina hiking trail

Top things to do
- Toast His Highness on 15 August – the only day the grounds of his castle in Vaduz open for visitors (and watch a magnificent firework display)
- Lunch among regal vines at Torkel, the ivy-clad royal restaurant in the capital
- Enjoy a concert in style at Balzers' classic 13th-century castle, Burg Gutenberg
- Hike the Fürstensteig, a rite of passage for every Liechtensteiner
- Skate along the Väluna Valley on cross-country skis

Getting under the skin
Read: David Beattie's *Liechtenstein: A Modern History* for the complete story on how this tiny country almost got wiped off the world map
Listen: to Vaduz-born classical composer Josef Gabriel Rheinberger (1839–1901)
Watch: movies beneath the stars during July's atmospheric Vaduz film festival
Eat: traditional local dishes like savoury *Käsknöpfle* (tiny cheese-flavoured flour dumplings) and sweet *Ribel* (a semolina dish served with sugar and fruit compote or jam)
Drink: local, rarely exported wine

In a word
Guten tag (Hello, good day)

Celebrate this
What did you do today? I hiked a whole country! It's not often you get to say that, but in diminutive Liechtenstein it's actually possible. So get those boots on and get walking.

Random fact
Liechtenstein is the only country in the world named after the people who purchased it.

1. Women wearing Liechtenstein's traditional costume

2. The alpine charm of Triesenberg, Liechtenstein's largest municipality

3. Schloss Vaduz is the official residence of the Prince of Liechtenstein's family, who purchased the castle in 1712

Liechtenstein

L — CAPITAL VADUZ // POPULATION 39,137 // AREA 160 SQ KM // OFFICIAL LANGUAGE GERMAN

With a history and monarchy as storybook as its mountain scenery speckled with tiled-roof stone castles and wintery snow scenes, Liechtenstein puts a whole new perspective on the European tour guide's 'doing a country'. Indeed, this tiny, wealthy nation landlocked between Alpine greats Austria and Switzerland can be 'done' in a day … or three at most. And what a sweet, toy-like experience it is. Its capital city, crowned with the king's castle and his vineyards, is of miniature proportions; mountains envelop two-thirds of the country, and a 25km stroll takes you from Liechtenstein's northern to southern tip.

L CAPITAL VILNIUS // POPULATION 2.7 MILLION // AREA 65,300 SQ KM // OFFICIAL LANGUAGE LITHUANIAN

Lithuania

Mother Nature has donated a decent dose of natural wonder to Lithuania, but you'll find that humans have left their stamp too, in undeniably weird and wonderful ways. White sandy beaches edge the Curonian Spit, an enchanting pig-tail of land dangling off the country's west coast, and thick forests guard twinkling lakes. The capital, Vilnius, is a beguiling artists' enclave, with mysterious courtyards, worn cobbled streets and crumbling corners overlooked by Baroque beauties. The country's oddities – among them a hill covered in crosses, a forest peopled with carvings of witches, and a sculpture park littered with Lenins – bring a flavour found nowhere else. Add a colourful history, and raw pagan roots fused with Catholic fervour, and you've got a country full of surprises.

Best time to visit
May to September

Top things to see
- Vilnius, the Baroque bombshell of the Baltics
- Thousands of crosses – some tiny, others gigantic – at the Hill of Crosses near northern town Šiauliai
- The slither of shifting sands that constitutes the remarkable Curonian Spit, with its mysterious Witches' Hill
- The red-brick Gothic castle of Trakai, in a fairy-tale lakeside location
- Hip second-city Kaunas with its art deco and street art, thumping nightlife and Museum of Devils

Top things to do
- Go fishing, boating, bathing and berry collecting in the country's beloved Lakeland
- Brave the winter and go ice fishing on the Curonian Spit
- Dunk yourself in the silky spa waters of Druskininkai
- Ponder the country's communist past at Grūtas Park
- Grab a beer with a sea view at port city Klaipėda

Getting under the skin
Read: *The Last Girl* by Stephan Collishaw, bringing Vilnius to life in a brilliant historical novel covering WWII to the 1990s
Listen: to avant-garde jazz from the Ganelin Trio; rock from Andrius Mamontovas, a household name in Lithuania for more than two decades; SKAMP, for hip hop and R&B
Watch: *Dievų Miškas* (Forest of the Gods), about a man imprisoned by both the Nazis and the Soviets
Eat: the formidable national dish of *cepelinai* (zeppelins), airship-shaped parcels of potato dough stuffed with cheese, meat and mushrooms, topped with a creamy sauce; save room (if you can) for *šimtalapis*, poppy seed cake
Drink: *midus* (mead: honey boiled with water, berries and spices, then fermented with hops); *stakliskes* (a honey liqueur); local beers Utenos and Švyturys; artisan coffee in Vilnius

In a word
Labas (Hello)

Celebrate this
Basketball is akin to religion in Lithuania, with Joniškis Basketball Museum as its temple – the worshipped national team enjoyed a victorious run, finishing in the top four for five successive Olympic Games.

Random fact
The country has a national perfume, the Scent of Lithuania, a mix of standard ingredients (bergamot, cedar, sandalwood) and the more unusual (moss and tree smoke).

1. Trakai Castle on Lake Galvė was first completed in 1409 and later restored by the Soviets in the 1950s

2. The Neris River flows through Vilnius; in winter temperatures rarely reach above freezing

3. In the 1960s, Lithuanians would plant a cross on the Hill of Crosses near Šiauliai in defiance of Soviet anti-religion edicts

CAPITAL LUXEMBOURG CITY // POPULATION 628,381 // AREA 2586 SQ KM // OFFICIAL LANGUAGES FRENCH, GERMAN & LUXEMBOURGISH

Luxembourg

The Grand Duchy of Luxembourg is a throwback to the days when Europe was a constantly evolving patchwork of tiny states. Sitting in the heart of Western Europe, its own heart is Luxembourg City, a Unesco-listed stunner commanding the confluence of the Alzette and Pétrusse rivers. To the north lies Gutland – the 'Goodland' of villages, pasture and the Moselle vineyards – and, beyond that, the Luxembourgish portion of the Ardennes, watered by numerous rivers and studded with perfect towns such as Esch-sur-Sûre and Vianden. Throughout this tiny country, ruled by Grand Duke Henri and his Cuban-born duchess, you'll find a people as proud and cosmopolitan as they are friendly.

Best time to visit
May to August, the sunniest months

Top things to see
- Modern art at Luxembourg City's Musée d'Art Moderne Grand-Duc Jean
- Superb views of the Old Town and river valleys from the Passerelle, Luxembourg City's dramatic 19th-century viaduct
- Woods and rock formations in the Müllerthal region, also known as Luxembourg's 'Little Switzerland'
- Echternach, a town steeped in Christian history and surrounded by forest
- Vianden's huge, ancient castle, rising imperiously above the town's cobbled streets

Top things to do
- Delve into the honeycomb innards of Luxembourg City's fortress casemates, dating back to 1644
- Stroll Luxembourg City's Old Town, then lunch alfresco on tree-lined Place d'Armes
- Mosey from one winery to another along the Route du Vin in the Moselle Valley
- Wander the five-tiered remains of the shattered Château Beaufort
- Visit the Unesco World Heritage–listed permanent exhibition by Luxembourg photographer Edward Steichen inside a castle in Clervaux

Getting under the skin
Read: *How to Remain What You Are*, a humorous look at Luxembourg ways by writer and psychologist George Müller
Listen: to classical music from the Luxembourg Philharmonic Orchestra; jazz from Gast Waltzing
Watch: *Lèif Lëtzebuerger* (Charlotte: A Royal at War) to catch Luxembourg's WWII history through the eyes of exiled grand duchess Charlotte
Eat: *quetschentaart*, an open tart, traditionally made with damson plums in autumn, now also made with other fruit and more widely available
Drink: a bubbly or fruity white Moselle wine bearing the quality label *'Marque Nationale du Vin Luxembourgeois'*

In a word
Moien (Hello, in Luxembourgish)

Celebrate this
In a bid to encourage an uptake in usage and a decrease in the number of traffic jams, all public transport in Luxembourg became free in March of 2020.

Random fact
Luxembourg's commitment to the pan-European ideal was confirmed in 1985 when the Schengen Agreement was concluded in the southern village of that name.

1. Looking across the Moselle River from Wincheringen; the river rises in France's Vosges mountains

2. Luxembourg's Philharmonie concert hall was designed by Christian de Portzamparc

3. The vines here grow mainly white grape varieties such as riesling, pinot gris and auxerrois

(M) **CAPITAL** ANTANANARIVO // **POPULATION** 27.0 MILLION // **AREA** 587,041 SQ KM // **OFFICIAL LANGUAGES** MALAGASY & FRENCH

Madagascar

Madagascar, one of the world's most famous islands, was not always just that, an island: as recently as 165 million years ago, it was still part of Africa. Seeded with life, it was eventually ripped off the continent and sent eastward into the Indian Ocean. There, the new island's plants and animals evolved in isolation, creating thousands of dumbfounding species. The remaining forests still teem with this outlandish life, such as dancing lemurs, and encounters can't help but make you gasp and giggle. The Malagasy people are relatively recent arrivals to the island. They believe family is central to life and their startling exhumation ceremonies prove the dead are just as important to them as the living.

Best time to visit
April to October (the dry season)

Top things to see
- *Tsingy*, surreal limestone pinnacles, rising into the sky in Parc National Bemaraha
- Allée des Baobabs, a wild dirt road lined by giant, ancient trees
- Modern Malagasy life in fast forward on the colourful streets of Antananarivo, and its past in stills within Musée de la Photo
- Lemurs within the lush cloud forests of Parc National de Ranomafana
- Aquatic life on the Great Reef near Anakao

Top things to do
- Experience the geological and biological wonders of Parc National de l'Isalo while hiking, cycling or swimming
- Walk in one of Madagascar's best primary rainforests in Parc National Masoala-Nosy Mangabe
- Listen to the wail of the rare indri, Madagascar's largest lemur, in Parc National Analamazaotra
- Step into an envy-evoking postcard at Andilana's beach on the island of Nosy Be
- Try not to get caught up in it all when walking through the remarkable 'spiny forest' in Parc National Andohahela, home to 12 lemur species

Getting under the skin
Read: *Beyond the Rice Fields* by Naivo, Madagascar's first novel to be translated into English; *A History of Madagascar* by Mervyn Brown
Listen: to *hira gasy*, live storytelling spectacles in Madagascar's central highlands
Watch: *Ady Gasy* (The Malagasy Way), a rich documentary of life in Madagascar by Nantenaina Lova
Eat: *vary hen'omby* (rice served with stewed or boiled zebu)
Drink: *rano vola* (rice water), a brown, smoky-tasting concoction

In a word
Manao ahoana ianao (How do you do?)

Celebrate this
Jean-Luc Raharimanana: a novelist, poet and playwright, who won the Jean-Joseph Rabearivelo

Poetry Prize by the age of 20; his works, often mixing stories of legends with current political affairs, are known for their portrayal of both the beauty of nature and the squalor of poverty.

Random fact
Rice is so significant in Malagasy culture that words used to explain the growth of it are the same as those used to describe a woman becoming pregnant and giving birth.

1. More than 110 species of lemur live only on Madagascar, including this critically endangered diademed sifaka and her young

2. The Allée des Baobabs in Morondava

3. Parc National des Tsingy de Bemaraha; *tsingy* translates as 'where one cannot walk barefoot'

M

CAPITAL LILONGWE // POPULATION 21.2 MILLION // AREA 118,484 SQ KM // OFFICIAL LANGUAGE ENGLISH

Malawi

The Great Rift Valley has long been associated with East Africa, yet some of its most spectacular geographical rewards are found in this southern African nation. Top of the list must be Lake Malawi, the lifeblood of the nation, which runs almost the length of the country and occupies a quarter of Malawi's total area. The rolling grasslands of the 2000m-high Nyika and Zomba plateaus and the rocky, waterfall-hewn massif of Mt Mulaje are all magnificent rift remnants too. This diverse landscape provides plenty of adventure activities, and recent wildlife conservation successes and burgeoning ecotourism have also made Malawi a rewarding safari destination. Malawians take pride in their nation's moniker 'The Warm Heart of Africa'.

Best time to visit
April to September

Top things to see
- The Big Five (lion, leopard, elephant, rhino, buffalo) in Majete Wildlife Reserve after recent conservation and re-introduction successes
- Beaches punctuated with baobabs and an unmatched view of Malawi's wild coast from Likoma Island
- More than a thousand colourful cichlid fish species in the clear waters of Lake Malawi
- New ecolodges near Livingstonia providing environmental and community benefits
- The beauty and wealth of life sustained by Lake Malawi's waters while on board the MV *Ilala* ferry, Malawi's grande dame of vessels

Top things to do
- Trek up the contorted massif of majestic Mt Mulanje for staggering vistas
- Spot elephants and hippos thriving along the Shire River within Liwonde National Park
- Kayak past crocodiles in the Bua River within Nkhotakota Wildlife Reserve
- Hike or ride a horse past zebras and antelopes in the magnificent highlands of Nyika National Park
- Swim, snorkel or simply float in the 'Lake of Stars' (aka Lake Malawi)

Getting under the skin
Read: *Soft Magic, Nectar* and *A Fire Like You*, books of poetry by Upile Chisala; *The Rainmaker*, a poetic drama by Steve Chimombo
Listen: to the Chichewa lyrics of Peter Mawanga & the Amaravi Movement
Watch: *The Boy Who Harnessed the Wind,* which follows an inventive schoolboy attempting to save his village; *The Road to Sunrise*, the story of two sex workers finding hope in each other
Eat: *nsima* (maize meal) and *chambo* (a fish from Lake Malawi)
Drink: Chibuku, a commercially produced local brew

In a word
Zikomo (Thank you)

Celebrate this
Upile Chisala: a Malawian poet who believes that her writing (and other women's) can create change by expressing their struggle for equality.

Random fact
Malawi is home to over 600 species of bird, and there are more fish species (over 1000) in Lake Malawi than in any other inland body of water in the world.

1. A performance of the Gule Wamkulu dance as the sun sets in Kasankha village

2. Elephants cool off in the Shire River, which flows into the Zambezi River in Mozambique

3. Looking across the Chiradzulu plains from Mt Mulanje

Best time to visit
May to September, for the least chance of rain

Top things to see
- The view over Kuala Lumpur's sea of skyscrapers from the Petronas Towers
- Tea plantations sprawling across the Cameron Highlands
- The dawn view from the summit of Mt Kinabalu, Malaysia's highest peak
- Colonial-era architecture and dragon-tiled clan-houses in George Town (Pulau Penang)
- Surreal acts of self-mortification during the Thaipusam festival at Batu Caves

Top things to do
- Get up close to the 'old man of the forest' in Borneo's Sepilok Orangutan Rehabilitation Centre
- Stay in an Iban longhouse on Sarawak's mighty Batang Rejang river
- Dive with sharks and turtles on the awesome reefs off Sipadan
- Enjoy the full tropical island experience at Pulau Perhentian or Pulau Langkawi
- Munch on *nasi lemak* (coconut rice steamed in banana leaves) in a traditional Melaka coffeeshop

Getting under the skin
Read: Tash Aw's *The Harmony Silk Factory* and Rani Manicka's *The Rice Mother* for two different takes on Malaysian multiculturalism
Listen: to the film-scores of P Ramlee; or the proud pop of Muslim icon Yuna
Watch: Yasmin Ahmad's award-winning human drama *Sepet*, or horror-comedy *Hantu Kak Limah*, showing Malaysia's kookier side
Eat: *roti canai* (fried flat bread with a rich curry dipping sauce) at one of Malaysia's 24-hour canteens
Drink: *teh tarik* ('pulled' tea with condensed milk); or *tuak* (rice wine from Borneo)

In a word
Malaysia boleh! (Malaysia can do it!)

Celebrate this
The traditional role of women in Malaysian society is being challenged by organisations such as Riding Pink, a female-run ride-hailing company that teaches women to drive, then employs them transporting female passengers.

Random fact
Malaysia is home to the largest flower in the world, the foul-smelling rafflesia, which can grow to more than a metre in diameter.

1. The Cameron Highlands offer an ideal environment for tea plantations

2. Malaysia is home to seven species of giant rafflesia, the flower with the world's largest bloom

3. George Town on the island of Penang has plenty of traditional 19th-century architecture

4. Kuala Lumpur's Petronas Towers have a sky bridge between the two towers at the 41st floor

Malaysia

M CAPITAL KUALA LUMPUR // POPULATION 32.7 MILLION // AREA 329,847 SQ KM // OFFICIAL LANGUAGE BAHASA MALAYSIA

Malaysia offers two vastly different territories for the price of one – Peninsular Malaysia, with its cities, forested highlands and coral-fringed islands, and Malaysian Borneo, whose jungles provide a haven for orangutans and indigenous communities. 'Unity in diversity' is the national motto of this famous melting pot, a blend of Malay, Indian, Chinese and European culture, with a serving of animist traditions courtesy of the Orang Asli (literally 'original people') of Sabah and Sarawak. The colonial period left a lasting impression, and political scandals continue to make the front pages, but modern Malaysia mostly exists in a state of harmony. For travellers, the focus is on the peninsula's beaches, the former colonial cities of George Town (Penang), Melaka and Kuala Lumpur, and the reefs and rainforests of Malaysian Borneo.

Best time to visit
December to April for fine weather, manta rays and whale sharks; or May to December for schooling hammerheads

Top things to see
- Sunrise over the surf from a palm-draped coral-sand beach
- The mesmerising underwater world just metres from your beach towel
- The mosques and bustling fish market in Male, the pocket-sized Maldivian capital
- Laid-back villages and the ruins of the British WWII air base on Gan island
- Whale sharks, manta rays and hammerheads performing a natural ballet beneath the surface of the Indian Ocean

Top things to do
- Dive or snorkel on spectacular *thilas* and *giris* (isolated reefs) and *kandus* (deepwater channels)
- See the Maldives from above on a scenic seaplane flight between the atolls
- Take a cruise to an outlying island on a *dhoni* (traditional Maldivian boat)
- Drop into a local cafe in Male for a snack-sized feast of *hedika* (short eats)
- Experience luxury at an underwater restaurant or day spa

Getting under the skin
Read: *Gatecrashing Paradise: Misadventure in the Real Maldives* by Tom Chesshyre – a budget traveller explores beyond the resorts
Listen: to *bodu beru* (big drum), the traditional folk music of the islands
Watch: *The Island President*, an award-winning documentary following the efforts of then-President Mohamed Nasheed to tackle rising sea levels caused by climate change
Eat: *garudia* (smoked-fish soup); or *hedika* – delicious, spicy fish-based snacks
Drink: *raa* (a sweet and tasty toddy tapped from the coconut palm)

In a word
Mabuti naman (I'm fine)

Celebrate this
Building on the Maldives' increasing awareness of climate change and rising sea levels, the country's newest resorts are focused on boosting biodiversity and reinforcing eco-aware and sustainable design and business practices.

Random fact
The highest point in the Maldives is just 2.4m above sea level – if sea levels continue to rise, plans are afoot to move the entire population to a new homeland overseas.

1. Male is one of the world's most densely populated cities

2. Giant oceanic manta rays live around the Maldives and can grow up to 5m in width

3. Biyadhoo on Kaafu Atoll is a scuba-diving paradise (and private resort)

M   CAPITAL MALE // POPULATION 540,544 // AREA 298 SQ KM // OFFICIAL LANGUAGE DIVEHI

Maldives

Floating just above the surface of the Indian Ocean, the islands of the Maldives are a glamorous playground for sun-seekers. This is where tourist brochure pictures come to life, and the scattered atolls are home to exclusive resorts, each on its own idyllic tropical island. The islands where ordinary Maldivians live were once off-limits, but the Muslim communities of the so-called 'inhabited islands' are increasingly opening up to visitors, and the natural beauty of the low-slung islands is matched only by the staggering richness of the coral reefs that lie between the atolls. Above the water though, there is sometimes trouble in paradise, thanks to the heavy-handed rule of the islands' autocratic government and the growing effects of climate change.

Best time to visit
October to February

Places of interest
- Djenné's Grande Mosquée, the largest (and most breathtaking) mud structure on the planet
- The legendary (yet humble) city of Timbuktu, with its historic manuscripts and mosques
- Bamako's live music venues where Mali's master musicians play
- The sleepy riverside town of Ségou with its *bogolan* (mud cloth) workshops
- The bustling port and salt-trading centre of Mopti, which rests at the confluence of the Niger and Bani Rivers
- Timeless Dogon villages at the foot of the dramatic Bandiagara Escarpment

Local customs
- Sharing and gift-giving are key elements of Malian society
- When nomadic Fulani women are not travelling by donkey caravans, they make handicrafts such as engraved gourds
- Tea, which forms a key part of the Tuareg culture's daily rhythm, is considered a friend of conversation
- The five-day Dogon Fête des Masques sees men wearing masks representing buffaloes, hyenas and Amma, the Dogon goddess of creation

Getting under the skin
Read: *The Bad-Ass Librarians of Timbuktu* by Joshua Hammer, the true story behind the efforts made to save Islamic manuscripts from Al Qaeda
Listen: to Tinariwen, Toumani Diabaté, Amadou and Mariam, the late Ali Farka Touré, Salif Keita, Oumou Sangaré, Rokia Traoré…the list is endless
Watch: *Mali Blues*, a documentary about four musicians standing in the face of jihadists
Eat: *tiguadege na* (meat in peanut sauce)
Drink: *bissap* or *djablani* juice (brewed from hibiscus petals); Beaufort (Malian beer)

In a word
Ere (peace, in Bambara)

Celebrate this
Abdel Kader Haidara and the other librarians of Timbuktu: they risked their lives to save some 350,000 priceless Islamic volumes from Al Qaeda militants in 2012.

Random fact
King Kankan Musa of the Mali Empire, believed to be the richest person in the history of the world, distributed so much gold en route to Mecca in the 14th century that it was a generation before the world's gold price recovered.

1. Each year, thousands of volunteers restore the Grande Mosquée of Djenné with mud at a festival

2. Masks are used by Dogon people during several rituals, including funerals

3. A Dogon village at Bandiagara Escarpment

CAPITAL BAMAKO // POPULATION 19.5 MILLION // AREA 1,240,192 SQ KM // OFFICIAL LANGUAGE FRENCH

Mali

More than a thousand years before Mali was Mali, it was at the heart of West Africa – and the now fabled city of Timbuktu, located on the southern edge of the Sahara, was one of the world's greatest centres of learning and a key trading post for its gold- and salt-rich empire. The city's legend and other compelling cultural attractions – Dogon Country, the mammoth mud-built Grande Mosquée of Djenné, the river port of Mopti, and the modern nation's enthralling music scene – later ensured the country became the darling of West African tourism in the 20th century. However, armed rebellions by Tuareg nomads, regular Islamic militant attacks (since the fall of Libya), and recent military coups have sadly made Mali off-limits to travellers for over a decade.

1. Jean de la Valette, who lends his name to this square, was a 16th-century Knight Hospitaller and the 49th Grand Master of the Order of Malta

2. The fortified city of Senglea was also known as Città Invicta (unconquered city)

3. Looking down Triq San Gwann in Valletta towards Fort St Angelo in Vittoriosa

CAPITAL VALLETTA // POPULATION 457,267 // AREA 316 SQ KM // OFFICIAL LANGUAGES MALTESE & ENGLISH

Malta

Sitting plumb in the middle of the Mediterranean, miniscule Malta has a strategic significance that belies its tiny size. Over the centuries, a succession of empires have squabbled over this rocky outpost to protect their fleets from rivals across the water. Today, the Maltese islands – there are actually three – are best known for beaches, nightlife, scorching summer sun and, increasingly, scuba diving. But scratch beneath the holiday-brochure gloss and you'll find tantalising glimpses of Europe's earliest civilisation, with rock-hewn temples carved centuries before the Egyptians even thought of their pyramids. Then there's the culture: a little bit British, a little bit Italian and a tiny bit Middle Eastern, tinged by memories of pirates, knights and sultans.

Best time to visit
April to June, September

Top things to see
- The fortified capital of Valletta, built by the crusading Knights of St John, with its new City Gate, Parliament Building and Opera House
- The view from the 'silent city' citadel of Mdina
- Malta's magnificent megalithic temples: Ġgantija, Ħaġar Qim and Mnajdra
- Marsaxlokk, a fishing village with seafood to die for

Top things to do
- Get lost in Roman, Maltese and British history at Vittoriosa's dazzling Maritime Museum
- Watch the curtain rise at Valletta's Manoel Theatre, one of Europe's oldest, dating to 1731
- Splash, swim and frolic in the Blue Lagoon
- Party like a Maltese during a *festa* – an infectious mix of music, food and fireworks

Getting under the skin
Read: British historian Ernle Bradford's *The Great Siege*, a gripping account of the epic 1565 battle between Ottoman Turks and the Knights of St John
Listen: to *ghana*, Maltese folk music that mixes the Sicilian ballad and the wail of an Arabic tune
Watch: *Simshar*, directed by Rebecca Cremona
Eat: a ricotta-stuffed *pastizza* (puff-pastry parcel); *aljotta* (garlic-spiked fish and tomato broth with rice); and *fenek* (rabbit) with spaghetti
Drink: a thirst-quenching rum and Kinnie (bitter orange and herb-flavoured soft drink)

In a word
Bongu (Hello)

Celebrate this
Malta has endured many conflicts. Facing starvation during the WWII siege, locals held firm and the whole island was awarded the George Cross by the British king in recognition of their heroism. The cross forms part of the Maltese flag.

Random fact
Malta's landscapes have featured in film and TV shows, including *Gladiator*, *Troy* and *Game of Thrones*.

(M) **CAPITAL** MAJURO // **POPULATION** 78,831 // **AREA** 181 SQ KM // **OFFICIAL LANGUAGES** MARSHALLESE & ENGLISH

Marshall Islands

It's a neon-blue water-world out in the Marshall Islands. This expanse of slender, flat coral atolls is so enveloped by tropical seas that anywhere at any time you can see, hear, smell and feel salty air and water. Marshall Islanders have embraced their remote environment to become some of the world's finest fisherfolk, navigators and canoe builders. Over the centuries, the British, Spanish, Germans, Japanese and Americans have all claimed these strategically located atolls. The US military presence remains huge and the traumatic effects of bomb testing still linger. The charm, however, lies in the country's outer islands, which retain the pristine feel of a Pacific paradise.

Best time to visit
The dry season from December to August in the south; the northern Marshalls are dry year-round and between September and November, rains can be a blessing

Top things to see
• A red-hot sunset from the Delap-Uliga-Darrit (DUD) lagoon
• Navigational stick charts, model *korkor* canoes and shell tools at the Alele Museum and Library
• The twisted wreckages of Japanese WWII Zeros, Betty bombers and more in the jungle foliage on Maloelap Atoll
• Beautiful, intricately woven mats, fans, baskets and 'kili bags' (once a favourite of Jackie Onassis) on sale

Top things to do
• Dive the WWII shipwrecks off Bikini Atoll
• Sail the waters of the Majuro lagoon in a *walap*, a traditional oceangoing canoe
• Catch a *boom-boom* (motorboat) for a *jambo* (trip) to one of the deserted outer islands
• Deep-sea fish off Longar Point on Arno Atoll
• Conquer your galeophobia and dive in the world's largest shark sanctuary

Getting under the skin
Read: *Stories from the Marshall Islands*, an anthology of traditional tales translated into English by Jack A Tobin
Listen: to the beat of *beet*, traditional Marshallese dance, influenced by Spanish folk music
Watch: *Jilel: The Calling of the Shell*, a Marshall Islands global-warming fairy tale
Eat: a snack of boiled, sweet pandanus fruit (just watch out for the hairy insides!)
Drink: ice-cold coconut water

In a word
Iakwe (Hello; literally 'You are a rainbow')

Celebrate this
Tiny Marshall Islands is home to the world's largest shark sanctuary. Shark fishing is prohibited in nearly 2 million sq km of Marshallese waters; fines collected from poachers go towards marine conservation projects.

Random fact
In 1946, Bikini Atoll was the site for the first peacetime detonation of an atomic bomb; the two-piece swimming costume (thought to be as awe-inspiring as the blasts) was named after the explosion site.

1. Jaluit Atoll is part of the Ralik chain of the Marshall Islands and covers about 11 sq km

2. Divers may encounter WWII American aircraft that were sunk after the war in the lagoon of Kwajalein Atoll

Best time to visit
November to March

Top things to see
- A colony of rare Mediterranean monk seals at Reserve Satellite du Cap Blanc
- Ben Amira: at 633m high, it's one of the world's largest natural rock monoliths
- Prolific birdlife on the Atlantic coast within Parc National du Banc d'Arguin
- Hundreds of colourful fishing boats beaching with their catches at Port de Pêche
- The dramatic old town of Ouadâne, built of brown stone

Top things to do
- Catch a lift on the iron-ore train (one of the world's longest) as it travels into the desert
- Clean up after time in the sand and enjoy some fine dining in Nouakchott
- Contemplate the past in Chinguetti, whose vestiges speak of an ancient Saharan city of Islamic learning
- Find shade beneath the palms of Terjît, one of the most verdant oases in the Sahara

Getting under the skin
Read: *The Desert and the Drum* by Mbarek Ould Beyrouk, winner of the Ahmadou-Kourouma Prize; Peter Hudson's *Travels in Mauritania*, which follows his exploration of the country on foot, donkey and camel
Listen: to Malouma, who has modernised traditional Moorish music
Watch: *Heremakono* (Waiting for Happiness) by Mauritanian director Abderrahmane Sissako, set in Nouâdhibou
Eat: at a *méchui* (traditional nomads' feast), where an entire lamb is roasted over a fire and stuffed with cooked rice
Drink: strong sweet tea; or the nomads' staple, *zrig* (curdled goat or camel milk)

In a word
Salaam aleikum (Hello, or Peace be with you)

Celebrate this
Biram Ould Abeid, the founder of the Initiative for the Resurgence of the Abolitionist Movement: he has spent his life fighting against slavery in Mauritania by using nonviolent tactics.

Random fact
Mauritania did not declare slavery a crime until 2007, making it the last country in the world to do so; however, tens of thousands of people are still thought to be enslaved here.

1. The libraries of Chinguetti, a medieval trading post, contain many Islamic texts, some dating from the Middle Ages; the city is a Unesco World Heritage Site

2. There's good surfing off Nouâdhibou, Mauritania's second city

3. Mauritania's 2km-long Train du Desert carries ore and passengers for 700km across the Sahara

M CAPITAL NOUAKCHOTT // POPULATION 4.0 MILLION // AREA 1,030,700 SQ KM // OFFICIAL LANGUAGE ARABIC (HASSANIYA)

Mauritania

Although flanked by the might of the cold Atlantic Ocean, Mauritania's greatest sea is the one made of sand. The Sahara's shifting dunes dominate this African country, flowing west from Algeria and Mali to the coast. The only interruptions are towering rock monoliths and remote oases, and some ancient Moorish towns like Chinguetti, which was a centre of Islamic learning from the 13th century. Mauritania is a traditional Islamic republic that is part Arab and part African, yet it has an identity all its own. Security issues in the Sahara have made much of eastern Mauritania out of bounds to visitors, but the western and coastal regions continue to welcome intrepid travellers and trans-Saharan overlanders moving between Europe and West Africa.

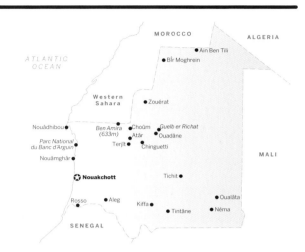

1. The St Denis River plummets more than 80m into the Black River Gorges at Chamarel Falls

2. Le Morne Brabant is dominated by a basalt monolith at the southwest tip of Mauritius

3. Browsing the central market of Port Louis is a sensory delight

M | CAPITAL PORT LOUIS // POPULATION 1.4 MILLION // AREA 2040 SQ KM // OFFICIAL LANGUAGE ENGLISH

Mauritius

Adrift in the Indian Ocean, this island paradise has collected many hangers-on since its fiery volcanic beginnings. One of the latest arrivals to wash up on its tropical shores was the human species. Today, Mauritius' social, gastronomic and architectural melange of Asian, African and European elements provides visitors with plenty to experience when not floating in azure seas, lounging on white-sand beaches or trekking through virgin rainforests. Minor tensions do exist between the Hindu majority, Muslims and Creoles, but respect and tolerance are intrinsic to Mauritian society. The arrival of people, whether passing by or permanent, caused the island's most famous endemic species to go the way of the dodo.

Best time to visit
May to November, for the driest skies

Top things to see
- Transcendent views over tropical lagoons, beaches and reefs from the top of Le Morne Brabant
- Remnants of rare coastal forests and endangered wildlife species on Île aux Aigrettes
- Eureka, a preserved Creole mansion, for an insight into the island's plantation past
- Trou d'Argent, an altogether staggering beach on the remote island of Rodrigues

Top things to do
- Immerse yourself at Rempart Serpent, La Passe St François and Colorado, three top dive sites
- Learn to kitesurf in Le Morne lagoon
- Open your eyes and nostrils to Sir Seewoosagur Ramgoolam Botanical Gardens
- Walk the trails within the Vallée de Ferney, home to rare and endemic birdlife
- Trek past waterfalls within the forested depths of the Black River Gorges National Park

Getting under the skin
Read: *The Last Brother* and *Blue Bay Palace* by Nathacha Appanah, which delves into local history
Listen: to Ti-Frère, the 'king of *séga*'; Keya, who fused *séga* and reggae to become the 'father of *seggae*'
Watch: *Lonbraz Kann* (Sugarcane Shadows) about the shutting of the island's sugar-cane factories
Eat: *rougaille* (a sauce of tomatoes, garlic and chilli with Creole-flavoured meat or fish)
Drink: *alouda* (an almond-based concoction best topped with ice cream)

In a word
Ayo! (a Creole expression for joy, sadness or surprise)

Celebrate this
Deegesh Maywah: a social entrepreneur who founded Employ Ability, which helps people living with disabilities enter and succeed in the workforce.

Random fact
A significant proportion of the country's electricity is produced from sugar cane.

CAPITAL MEXICO CITY // POPULATION 128.65 MILLION // AREA 1.96 MILLION SQ KM // OFFICIAL LANGUAGE SPANISH

Mexico

Mexico's roll call of famous sights and activities reads like a list of old friends: Mexico City museums; scuba diving around Cancún and Playa del Carmen; surfing and whale-watching in Baja; Chichén Itzá's Maya monuments; boisterous celebrations for the Day of the Dead. But there are plenty of surprises once you get past the touristy top draws – jungles, deserts, canyons, snow-capped mountains, colonial cities and indigenous culture. In recent years, Mexico's reputation as America's favourite escape has taken a knock from drug cartel violence, but travel patterns have flexed to avoid the hotspots, and the old charm still endures. Forget what you think you know, and you'll find more than you expected in Mexico.

Best time to visit
October to May, to avoid extreme temperatures

Top things to see
- Mexico's megacity capital: vast and chaotic but also cultured and cool
- Maya temples rising above the dense jungle at Palenque
- Stately old-world charms in colonial Oaxaca and blissful beaches on the Pacific coast
- The vast abyss of the Copper Canyon, snaking through the Sierra Madre
- Classy beach life in Playa del Carmen

Top things to do
- Climb the magnificent, mysterious pyramids of the sun and moon at Teotihuacán
- Go on the ultimate Mexican road trip through Baja's wild and desolate interior
- Sample real Mexican cuisine at a local *cantina*
- Admire Mexico's colonial legacy in Unesco-listed Guanajuato
- Feel the breeze of a billion butterfly wings at the winter refuge of Reserva Mariposa Monarca

Getting under the skin
Read: *Where the Air is Clear* by the great Carlos Fuentes; *Frida Kahlo and Diego Rivera* by Isabel Alcántara and Sandra Egnolff, the story of the two cultural icons
Listen: to some traditional mariachi music; Los Tigres del Norte and Cafe Tacuba – pioneers of *rock en español*
Watch: a tale of near-hallucinatory romance in *Like Water for Chocolate*; a mesmerising central performance from Yalitza Aparicio in *Roma*
Eat: the ubiquitous tortilla: soft and stuffed as burritos, enchiladas and quesadillas, or crisp and filled as tacos, taquitos and tostadas
Drink: *jugos naturales*, especially the bloodlike *vampiro* fruit juice (beetroot and carrot); the three spirits distilled from the agave plant: tequila, mezcal and pulque

In a word
¿Que onda? (How's it going?)

Celebrate this
A self-taught artist, Frida Kahlo drew inspiration from the country she lived in and her many painful personal experiences to create works that are both uniquely Mexican while also speaking to a global audience.

Random fact
Mexico City is sinking by up to 40cm per year as its burgeoning population drains water from the lake on which the city is built.

1. Looking towards Parroquia de San Miguel Arcángel in San Miguel de Allende, Guanajuato

2. Spot whales in Loreto Marine Park in Baja California

3. Restoring the Mayan ruins at Palenque in Chiapas

M

①

②

Best time to visit
October to March, when trade winds provide relief from FSM's energy-sapping humidity

Top things to see
- Yap's enormous stone money doughnuts, an ancient form of currency
- The ramshackle, pulsing hub of Kolonia, FSM's biggest town
- Hardworking artisans and local dance performances on Yap island
- The 'Venice of Micronesia', Pohnpei's ancient stone city, Nan Madol

Top things to do
- Dive Chuuk's veritable museum of Japanese WWII wrecks and spot Yap's graceful manta rays
- Imagine ancient life in the Lelu Ruins secreted in dense tropical vegetation on Kosrae
- Hike Pohnpei's and Kosrae's lush, rugged volcanic terrain
- Kayak through Kosrae's magical mangroves
- Watch the brilliant sunset behind the Faichuk Islands from Weno on Chuuk

Getting under the skin
Read: Emelihter Kihleng's *My Urohs*, the evocative first-ever collection of poetry in English by a Micronesian writer
Listen: to the compilation *Spirit of Micronesia*, recordings of traditional songs from around the region
Watch: *Globe Trekker's* Megan McCormick island-hop through Pohnpei, Yap and Chuuk
Eat: or rather chew *buw* (betel nut) with the locals and stain your mouth red
Drink: *sakau* (local narcotic kava drink made from the roots of pepper shrubs)

In a word
Fager (Yapese), *Kompoakepai* (Pohnpeian), *Pwipwi* (Chuukese), *Kawuk* (Kosraean) – the word 'friend' from the respective islands

Celebrate this
The 'Micronesia Challenge' – a pledge to protect huge swathes of FSM's marine territory – has rescued hundreds of fish and coral species; it's now aiming to establish the largest protected marine area in the region.

Random fact
Operation Christmas Drop, the world's longest-running humanitarian airlift, has seen the US Air Force drop boxes of supplies and gifts on FSM annually since 1952.

1. Ant Atoll lies off the west coast of Pohnpei and is famed for its diving and snorkelling

2. There's no such thing as small change on the island of Yap; this is the local currency of stone coins

3. Eagle rays fly through the ocean around Micronesia

M **CAPITAL** PALIKIR // **POPULATION** 102,436 // **AREA** 702 SQ KM // **OFFICIAL LANGUAGE** ENGLISH

Micronesia, Federated States of

And now for somewhere completely different. The Federated States of Micronesia (FSM) is made up of four unique and otherwise unrelated island states: Kosrae, Pohnpei, Chuuk and Yap. Each region has cultures and traditions as colourful, distinct and diverse as the fish and coral formations that paint the fringing reefs. Kosrae is a Pacific paradise and arguably FSM's most beautiful island; Pohnpei is home to mysterious ancient ruins and a plethora of landforms; Chuuk is renowned for its wreck diving; and Yap is a fiercely traditional state retaining a true island spirit. If you can't find something in the diversity of FSM to expand your view of the world, check your pulse.

1. The Presidential Palace of Chișinău was built under Soviet rule in the 1980s; Moldova declared its independence in 1991

2. Looking towards one of several religious sites in Orhei National Park

3. Grapes harvested in the south of Moldova make some of Europe's best-value wines

M CAPITAL CHIŞINĂU // POPULATION 3.3 MILLION // AREA 33,851 SQ KM // OFFICIAL LANGUAGE MOLDOVAN

Moldova

Cradled deep within Eastern Europe, corralled from the sea by neighbouring countries, and strewn with rippling hills, Moldova remains one of the continent's least touristed destinations. Sharing much, linguistically and culturally, with its neighbour Romania, it also contains two semi-autonomous regions: the largely Slavic Transdniestr, bordering Ukraine, and Gagauzia in the south, the home of the Turkic Gagauz. While it's most famous for the precious output of its extensive vineyards, there's much more to Moldova than quality plonk, including the cafe scene and parks of the capital, Chişinău, the 13th-century monastic complex dug into the caves of Orheiul Vechi, and the bird-laden Prutul de Jos Biosphere Reserve.

Best time to visit
May to September

Top things to see
- Cathedral Park, framed by Chişinău's own Arc de Triomphe, flower market and Orthodox Cathedral
- Caves dug by 13th-century monks at Orheiul Vechi Monastery Complex
- The Soroca Fortress on the Dniestr, a behemoth from the late 15th century and the reign of Moldavian Prince Stefan cel Mare
- The Pushkin Museum in Chişinău, where the great writer was exiled and wrote some of his novels

Top things to do
- Browse the world's largest wine collection in over 200km of subterranean cellars at Mileştii Mici
- Go on a kayaking expedition on the Dniestr River or bike through beech forests in the Plaiul Fagului Nature Reserve
- Travel back in time to Tiraspol, capital of Transdniestr and a lonely vestige of communism

Getting under the skin
Read: *The Good Life Elsewhere* by Vladimir Lorchenkov, a satire about Moldovan villagers trying to emigrate to Italy
Listen: to the unique 'bang and boom' sound of folk/hip hop/punk innovators Zdob şi Zdub
Watch: Nap Toader's *Wedding in Bessarabia*, a comedy about the Moldovan-Romanian culture clash
Eat: *pelmeni* (Russian-style ravioli); or *sorpa* (a spicy mutton soup made by the Turkic Gagauz)
Drink: local wine varietals including Fetească and Rara Neagra; or traditionally distilled vodka

In a word
Buna (Hello)

Celebrate this
The capital of Gagauzia autonomous region, Comrat, is home to a museum dedicated to the Gagauz, Moldova's little-known Turkic minority.

Random fact
Cognac from Tiraspol's Kvint distillery was allegedly the favourite tipple of Soviet cosmonaut Yuri Gagarin.

KEREN SU | GETTY IMAGES // ALEXANDER SMALAR | GETTY IMAGES // MOUNTAINS HUNTER | SHUTTERSTOCK

POPULATION 39,000 // **AREA** 2 SQ KM // **OFFICIAL LANGUAGE** FRENCH

Monaco

Prince Albert II is monarch of this dynastic fairy tale, entirely surrounded by France on Europe's most celebrity-studded coastline – the Côte d'Azur, or French Riviera. Glitzy, glamorous and bubbling over with self-assurance, this tiny seaside state sizzles with resident millionaires, luxury yachts and day trippers by the millions who traipse up to 'the Rock' to see the prince's palace, sip Champagne at Café de Paris and try their luck at the casino. Monégasque life is the high life, all-embracing the second you step off the train into Monaco's swanky, subterranean, marble-clad station. Above ground, skyscrapers jostle for sunlight as billion-dollar plans to expand Dubai-style out to sea hover on this unusual country's glittering blue horizon.

Best time to visit
April to June, September, October

Top things to see
- The changing of the guard at the Palais du Prince
- Sharks above your head at the Musée Océanographique
- The graves of Grace Kelly and her Monégasque Prince Charming, Rainier III, in the cathedral
- Billionaire yachts moored at the port
- World-class racing drivers tearing around during the Formula 1 Grand Prix in May

Top things to do
- Risk a lot (or a little) amid belle époque opulence at Monte Carlo Casino
- Enjoy a caviar body treatment at the Thermes Marins Monte Carlo
- Lunch at Le Louis XIV, Monaco's most prestigious dining address
- Hike near century-old cacti in the Jardin Éxotique
- Motor into France between sea and cliff along the hair-raising Grande Corniche

Getting under the skin
Read: Graham Greene's *Loser Takes All*, about a couple honeymooning in 1950s Monte Carlo
Listen: to 1930s music-hall singer Charles Colbert sing *The Man Who Broke the Bank at Monte Carlo*
Watch: the future Princess of Monaco, Grace Kelly, and Cary Grant in Hitchcock's Monaco/Riviera-set *To Catch a Thief*
Eat: Monégasque specialities *barbajuan* (spinach and cheese pasty) and *stocafi* (dried cod in a tomato sauce)
Drink: Champagne aperitifs at waterfront nightclubs

In a word
Bon giurnu (Hello, in Monégasque)

Celebrate this
Portier Cove is Monaco's first land reclamation scheme with green credentials, building new (luxury) homes while safeguarding the natural environment, using renewable energy and showcasing architecture by the likes of Renzo Piano.

Random fact
Although famous for its casinos, Monaco forbids its citizens from even entering the gambling rooms (employees excepted).

1. The glitzy interior of Monte Carlo Casino

2. Real estate in La Condamine around Monaco's harbour is some of the world's priciest

3. Spot beautiful boats in Monaco's harbour

Best time to visit

June to September, avoiding the cold of deep winter

Top things to see

- The Naadam festival, a vigorous annual contest of horse races, archery and traditional wrestling
- The clarity of spectacular Khövsgöl Nuur lake
- Altai Tavan Bogd National Park, with its snow-peaks, sparkling lakes and Kazakh eagle hunters
- The ruined Mongol capital of Karakorum and the nearby Tibetan Buddhist monastery of Erdene Zuu Khiid
- Ulaanbaatar, one of the world's most curious capital cities, where modern and nomadic lifestyles collide

Top things to do

- Channel the spirit of Genghis Khan, and gallop on horseback across endless open grasslands
- Spot gazelles, argali sheep and ibex in the country's epic wildernesses
- Trek with Bactrian camels through the silent Gobi Desert
- Overnight in a *ger*, and be treated like an honoured guest by your herder hosts
- Gaze up at the incredible ceiling of stars almost anywhere in the country

Getting under the skin

Read: *Moron to Moron: Two Men, Two Bikes, One Mongolian Misadventure* by Tom Doig; or Louisa Waugh's travelogue of nomad life, *Hearing Birds Fly*
Listen: to the otherworldly sound of *khöömii* (Mongolian or Tuvan throat singing)
Watch: Sergei Bodrov's *Mongol*, a biopic of Genghis Khan; or Erdenebileg Ganbold's historical drama, *The Steed*, set against a backdrop of revolution
Eat: *boodog* (barbecued roast marmot, cooked from the inside with hot stones and crisped to perfection with a blow torch)
Drink: *airag* (fermented mare's milk); or *suutei tsai* (salty milk tea)

In a word

Sain baina uu (Hello)

Celebrate this

Founded by four female lawyers, Women for Change is pushing back social boundaries in Mongolia; recent projects include patriarchy-challenging street art, an anti-domestic-violence graphic novel, and nude recreations of classic paintings to empower women with body confidence.

Random fact

Mongolia is the least densely populated country on Earth – just two people per square kilometre.

1. Life in a Mongolian *ger* is not without its creature comforts

2. Camel caravans cross the Badain Jaran desert

3. One man and his eagle at Ölgii Eagle Festival, which is held every October and grew in popularity after the 2016 film *The Eagle Huntress*

4. There are few roads in Mongolia and only 20% are paved

M CAPITAL ULAANBAATAR // POPULATION 3.2 MILLION // AREA 1,564,116 SQ KM // OFFICIAL LANGUAGE MONGOLIAN

Mongolia

Sitting serenely under what might be the world's biggest skies, Mongolia is the perfect place for anyone who values personal space. Almost devoid of fences, roads or even towns, this vast sweep of undulating steppes and rounded mountain ranges is best known for the conquering rampages of Genghis Khan, but in the modern age, Mongolia offers a kind of emptiness and silence found almost nowhere else. Travelling by horse (or its modern equivalent, the 4WD), visitors can step into a way of life that has changed little since the days of the Silk Road, staying with nomadic herders and sleeping under dark skies, beneath a canopy of felt and animal skins in a traditional *ger* (yurt). It's a beacon for those seeking untouched, vast landscapes and a life-affirming sense of shared humanity.

M

Montenegro

Montenegro (Crna Gora) may have remained unknown to the West for centuries, but it was certainly well known to Adriatic pirates, Venetian plunderers, Ottoman pashas and Yugoslav technocrats. The unforgiving landscape and its people resisted all comers, emerging as an independent nation in 2006. It is a country that prides itself on the deep-rooted code of 'humanity and bravery' and one that comfortably accommodates Orthodox monasteries, Albanian mosques and communal promenading every evening. As suggested by its brooding name (Crna Gora means 'Black Mountain'), Montenegro is a high, rugged place. Glacial lakes and dizzying canyons are the showpieces of the northern mountains; on the coast, flaming pomegranate trees and the azure waters of the Adriatic have an equally fierce beauty.

Best time to visit
April to September

Top things to see
- Almighty panoramas over Kotor and the cyan sea from St John's Hill
- The historic capital of Cetinje, an idiosyncratic mix of grand city and cosy village
- Ostrog Monastery, dramatically carved into a cliff and the spiritual heart of Montenegro
- The mausoleum of national hero, Petar II Petrović-Njegoš, in Lovćen National Park
- The tiny, yet super-photogenic island of Sveti Stefan

Top things to do
- Hang on for the white-knuckle drive from Kotor to Cetinje
- Feel the rush on a rafting trip down the Tara River
- Cool off in the limpid waters of the Adriatic at the beaches around Budva
- Peer through your binoculars at Lake Skadar, one of Europe's most important wetlands
- Hike in Durmitor National Park, home to eagles, bears, wolves and glacial lakes

Getting under the skin
Read: *Catherine the Great and the Small* by Olja Knežević, a cross-generational story of family, love and Balkan life
Listen: to epic poetry sung to the accompaniment of the *gusle* (vaguely akin to a single-stringed sitar)
Watch: *Packing the Monkeys, Again!*, the 2004 debut of Montenegro's first female film director Marija Perović
Eat: *lignje na žaru* (grilled squid) on the coast; or, in the mountains, *jagnjetina ispod sača* (lamb baked in a metal pot under hot embers)
Drink: *loza* (grape brandy); or Nikšićko, the local beer

In a word
Dobro došli (Welcome)

Celebrate this
In Montenegro, farm-to-table cuisine is not a fad but a way of life. Many events celebrate delicious local produce, such as the highlands' Days of Pljevlja Cheese in October and Plav's Days of Blueberries in July, or Lake Skadar's Festival of Wine and Bleak in December.

Random fact
More than half of Montenegro is 1000m above sea level, and 15% is higher than 1500m.

1. Sveti Stefan is a small, hyper-exclusive island-resort but some of the beaches are open to non-residents

2. It's easy to explore the Bay of Kotor on the surrounding roads

3. Lake Skadar straddles the border of Albania and Montenegro, where it is a national park

Best time to visit
October to April

Top things to see
- The rainbow of dye pits in the medina in Fez, the most intact medieval Arab city in the world
- Erg Chigaga, the largest of the Sahara Desert's sand seas in Morocco
- Sunset lighting up the many blue hues of the quiet mountain town of Chefchaouen
- The palm oases and red cliffs of Dadès Gorge
- European and African cultures balancing beautifully in the port of Tangier

Top things to do
- Wade into the crowds after sunset to dine at one of the stalls in Marrakesh's Djemaa el-Fna square
- Trek deep into the Atlas Mountains with mules, sleeping at local Berber homestays
- Harness the famed coastal winds – *alizee*, or *taros* in Berber – while kitesurfing in Essaouira
- Admire the scale of Heri Es Souani, the ruins of the king's stables and granary outside the imperial city of Meknès
- Haggle in the souqs until you're knee-deep in colourful *balgah* (heelless leather slippers)

Getting under the skin
Read: *The Voices of Marrakesh* by Nobel laureate Elias Canetti; *Hope and Other Dangerous Pursuits* by Laila Lalami, which follows the lives of four Moroccans heading for Spain in a raft
Listen: to the rhythms of the Master Musicians of Joujouka; Mehdi Nassouli, who blends classic *gnawa* (Sufi-inspired music) with hip hop and other genres
Watch: *Adam*, the story of an unwed pregnant mother seeking refuge from a widow
Eat: *seksu* (couscous) – steamed for hours and heaped with meat or vegetables
Drink: green tea with mint

In a word
Lebas? (How are you?)

Celebrate this
Astronomer, explorer, astrophysicist and Arab Woman of the Year (2015) Merieme Chadid, who set up an astronomical observatory in Antarctica.

Random fact
Casablanca's Hassan II Mosque accommodates 25,000 worshippers at prayer, and its minaret is 210m high – the tallest in the world.

1. Losing yourself among the stalls and sales pitches in the souq of Marrakesh

2. There are as many reasons suggested as to why Chefchaouen was painted blue as there are hues

3. Roses are harvested by the Berber women of H'dida village in the M'Gouna Valley

4. Ait Ben Haddou: this Unesco-protected red mudbrick *ksar* (fortified vilage) still resembles its days in the 11th century as an Almoravid caravanserai

M CAPITAL RABAT // POPULATION 35.6 MILLION // AREA 446,550 SQ KM // OFFICIAL LANGUAGES ARABIC & BERBER

Morocco

From Near Eastern nomads and Phoenicians to Romans and Berbers, Morocco's long history has been shaped by numerous cultures. Evidence of all is seen today, in bloodlines, traditions, cuisine, architecture and ancient monuments. The nation's cities also speak to this past, but to the future too, with the historic medinas of Fez, Marrakesh and Essaouira contrasting with the urban glass and glitz seen in Casablanca, Tangier and Rabat. The landscapes that carpet this slice of North Africa are just as varied, much like the richly coloured and patterned rugs for sale in the atmospheric souqs. There are the mountains of the High Atlas and the Rif, the dunes and oases of the Sahara, and the staggering coastlines of the Atlantic and Mediterranean.

Best time to visit
May to November, during the cooler, drier season

Top things to see
- Centuries of history on show in the narrow, atmospheric lanes of Mozambique Island
- The azure waters of the Indian ocean gently kissing the white sands of the Quirimbas Archipelago
- Lions, elephants and birdlife flourishing in the forests, canyons, savannahs and wetlands of Gorongosa National Park
- The quiet streets and relaxed beaches of Inhambane

Top things to do
- Snorkel or dive to see dugongs, corals and colourful fish in the waters of the Bazaruto Archipelago
- Hike over precipitous log bridges and through rushing mountain streams in the cool Chimanimani Mountains
- Wind your way through the heart of rural Africa on the Nampula–Cuamba train
- Explore Lake Niassa (aka Lake Malawi), with its clear waters, secluded coves and star-filled skies
- Sail on a traditional dhow past deserted islands and remote beaches

Getting under the skin
Read: Mia Couto's novels *Confessions of a Lioness* and *The Last Flight of the Flamingo*, both of which provide a window into the nation's soul
Listen: to Mabulu, a Maputo-based collective that fuses modern dance music with traditional sounds and social-conscious lyrics or Gabriela's Afro-pop
Watch: *Sleepwalking Land*, the story of an orphaned refugee during the civil war
Eat: *matapa* (cassava leaves with peanut sauce)
Drink: a Raiz beer – cold, if you can find it

In a word
Paciência (Patience)

Celebrate this
Its mountainous passes were once a frontline of war, but now the Chimanimani region is one of the world's newest national parks – it features dense forests, ancient San rock paintings and wildlife populations recovering from decades of poaching.

Random fact
Mozambique's national flag is the only one in the world to feature a modern weapon.

1. An anemone dancing in the current around beautiful Bazaruto Archipelago, just off Mozambique's coast

2. Mozambican Afro-pop star Gabriela in the recording studio

3. The capital's Independence Square: in 1976, a year after independence, President Samora Machel changed the city's name to Maputo, honouring an early chief who had resisted Portuguese colonisation

CAPITAL MAPUTO // POPULATION 30.1 MILLION // AREA 799,380 SQ KM // OFFICIAL LANGUAGE PORTUGUESE

Mozambique

With its national flag featuring an AK-47 assault rifle, Mozambique wears its troubles on its sleeve. Not long ago the nation basked in its glowing image as a post-war success story, with a fresh peace deal, a landmine-free declaration and a blossoming ecotourism industry. Yet the discovery of massive natural gas reserves off its coast has triggered some violent historical divisions, which have in turn opened the door for Islamic extremism in the nation's north. When harmony reigns, the nation is a sight to behold – dune-backed beaches line the mainland, azure waters around remote archipelagos thrive with aquatic life and wildlife roams its forests and savannahs. And its enigmatic fusion of African, Arabic, Indian and Portuguese influences provides intoxicating cultural experiences.

(M) **CAPITAL** NAY PYI TAW // **POPULATION** 56.6 MILLION // **AREA** 676,578 SQ KM // **OFFICIAL LANGUAGE** BURMESE

Myanmar

Long shunned as a pariah state because of its repressive military government, the nation formerly known as Burma enjoyed a ray of optimism that followed the landslide victory of Aung San Suu Kyi's National League for Democracy in 2015. That was quickly undermined by the persecution of Rohingya people in the west of the country and then a military coup in 2021, restoring Myanmar's rogue state status. That's one side to Myanmar. Juxtaposed against this is the Myanmar travellers actually experience: a land of rich culture and dense layers of history, where golden stupas adorn the landscape like giant candelabras, and the pace of life is dictated by the flow of the mighty Ayeyarwady River. Few leave unmoved by the experience of visiting, but human rights abuses keep Myanmar in the news for all the wrong reasons.

Best time to visit
November to February (the cool season)

Top things to see
- The breathtaking plain of temples at Bagan, ransacked to ruin by Kublai Khan
- Wandering through a fantasy garden of pagodas at the jewel-encrusted Shwedagon Paya
- Wonky bridges and hilltop stupas in the villages dotted around historic Mandalay
- Foot-rowers swirling through the morning mist on Inle Lake
- The timeless pace of village life in sleepy Hsipaw

Top things to do
- Cruise along the Ayeyarwady to Bagan on a nostalgically old-fashioned passenger boat
- Trek to deeply traditional tribal villages around Kalaw or Kengtung
- Splash through the surf on idyllic Ngapali Beach
- Climb Mt Popa to commune with the *nats* (animist spirits)
- Shop for puppets and Buddhist amulets at Yangon's Bogyoke Aung San Market

Getting under the skin
Read: Pascal Khoo Thwe's memoir *From the Land of Green Ghosts*; or George Orwell's *Burmese Days*
Listen: to the rousing rock of Lay Phyu and Iron Cross, the slick pop of Sai Sai, or the activist punk rock of Rebel Riot
Watch: The Maw Naing's *The Monk*, one of Myanmar's first independent features; or historical romp *The Great Myanmar* by Jae Sung Jeony and Aung Kyaw Moe
Eat: at a traditional Burmese canteen – a typical meal includes *htamin* (rice), *hin* (curries), *thoke* (salads), *peh-hin-ye* (lentil soup) and *balachaung* (fiery shrimp and chilli paste)
Drink: Dagon Beer, Myanmar Beer or Spirulina Beer, which claims anti-ageing properties

In a word
Mingalaba (We are blessed)

Celebrate this
Rohingya activist Wai Wai Nu: she has garnered worldwide recognition for her work promoting human rights and equality and raising awareness of the treatment of marginalised peoples beyond Myanmar's borders.

Random fact
Myanmar is one of only three nations to eschew the metric system; locally, weights are measured in *viss* (equivalent to 1.6kg), *tical* (16g) and a series of smaller divisions.

1. Monks pedal past a Buddha at Yan Aung Nan Aung Hsu Taung Pyi Pagoda near Inle Lake

2. Bagan's temples punctuate a 26-sq-mile area and were at the heart of an ancient kingdom

3. Village life in Pane Ne Pin in Shan state, which borders China

N CAPITAL WINDHOEK // POPULATION 2.6 MILLION // AREA 824,292 SQ KM // OFFICIAL LANGUAGE ENGLISH

Namibia

Nowhere in the world is such visual splendour matched with such unmerciful harshness – if anything is to survive in Namibia, in the vast sea of dunes, within the narrow rocky canyons or along the barren Skeleton Coast, it has to be equally extraordinary. Thankfully, nature has provided just such a cast. Its members include desert-adapted elephants, rhinos and lions, as well as fog-basking beetles, rare mountain zebras and cartwheeling spiders. Its people too are some of the most resilient around, although they are few and far between, making Namibia the least densely populated nation on the continent. As such, its wide-open spaces and safe road network ensure it is the best place in Africa for a road trip or self-drive safari.

Best time to visit
May to September (the dry season)

Top things to see
- The heavens and Milky Way coming to life at night in Africa's only International Dark-Sky Reserve, the NamibRand
- Jackals, bleached whale bones and hulking shipwrecks along the desolate Skeleton Coast's mist-shrouded shore
- The earth opening up before you as you approach the edge of the Fish River Canyon
- Sunrise casting shadows across fiery red dunes at Sossusvlei in Namib-Naukluft National Park
- The Namib's sands delightfully digesting the ghost town of Kolmanskop

Top things to do
- Observe some of Africa's most prolific wildlife while on safari in Etosha National Park
- Camp in the surreal boulder-strewn landscape beneath Spitzkoppe, the 'Matterhorn of Africa'
- Cruise the Caprivi Strip and embrace the wet and wild life on the banks of the Chobe River
- Track endangered black rhinos through the parched landscape of the remote Kaokoveld
- Admire the ancient petroglyphs of the San people at Twyfelfontein

Getting under the skin
Read: *Born of the Sun*, which follows the coming of age and political awakening of author Joseph Diescho; *The Sheltering Desert*, a non-fiction tale of two Germans evading internment during WWII
Listen: to Gazza, who mixes *kwaito* with reggae, dancehall and hip-hop beats
Watch: *Hairareb*, a story that unfolds when a boy finds his father's diary; *Namibia: The Struggle for Liberation*, the story of Namibia's independence told through its first president
Eat: *oshiwambo* (a mixture of spinach and beef)
Drink: prickly-pear-cactus schnapps

In a word
Goegaandit?/Matisa?/Kora? (How are you? in Afrikaans/Damara/Herero)

Celebrate this
Stigmatised in her community at the age of 17 when she fell pregnant and was diagnosed with HIV, single mother Livey van Wyk went on become her city's (and the nation's) youngest mayor less than a decade later.

Random fact
Namibia is home to the world's largest non-subglacial underground lake; named Dragon's Breath Cave.

1. Fish River Canyon is Africa's largest canyon and more than 550m deep in places with popular hiking trails

2. Seals congregate on Namibia's Skeleton Coast to fish in the Benguela Current, attracting orcas

3. Members of the pastoralist Himba tribe of northern Namibia work on their tools

4. The sands of Sossusvlei probably originated in the Kalahari between three and five million years ago

Best time to visit
Hot and humid year-round; it's best to visit from March to October when it's driest

Top things to see
- The eerie rusted cantilever cranes used to load phosphate for export
- WWII relics scattered around the island, including a Japanese jail block on Command Ridge
- A game of Australian Rules football or locals weightlifting – Nauru's two national sports
- *Ekawada*, string figures stretched across the hands like 'cat's cradle' to tell traditional stories
- The ruins of the once-splendid presidential palace, burned down by a local mob in 2001

Top things to do
- Walk the pinnacled remnants of the now defunct phosphate mines
- Catch, tag and release marlin, yellow-fin tuna, barracuda and more with the island's fishermen
- Quaff a cold beer while watching a fiery South Seas sunset over coconut palms
- Swim and snorkel off the beach by the Menen Hotel

Getting under the skin
Read: *The Undesirables: Inside Nauru* by Mark Issacs, a damning look inside the island's detention centre, where the author worked with asylum seekers
Listen: to the strange *kik-kirrik* cry of the noddy bird
Watch: *Nauru – Paradise Ruined*, the 2011 documentary about Nauru's rise and fall
Eat: a fresh seafood barbecue
Drink: *demangi*, the island's take on fermented toddy made from coconut palm sap

In a word
Kewen (Gone, dead)

Celebrate this
There's not much to recommend it to tourism, but if you really don't like crowds, Nauru could be for you: fewer than 200 people visit this flat little island each year.

Random fact
During the phosphate boom in the 1980s, Nauru was the second-richest country in the world in terms of per-capita income; 40 years later, it's one of the world's top five poorest.

1. Feeding (and catching) frigatebirds is an age-old tradition on Nauru

2. Phosphate was once Nauru's main export, now derelict loaders dot the island

3. Beaches on this small, remote island often feature volcanic rocks

N · **CAPITAL** YAREN (DE FACTO) // **POPULATION** 11,000 // **AREA** 21 SQ KM // **OFFICIAL LANGUAGES** NAURUAN & ENGLISH

Nauru

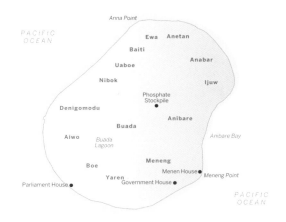

Nauru's limited beauty can be glimpsed along its coast: seabirds swoop over green cliffs beside wild seas, and sunsets are nothing short of spectacular. Head to the tiny island's interior however, and you'll find deforestation from phosphate mining and an eerie landscape of limestone pinnacles. The exposed rock reflects the sun's rays and chases away the clouds so there's lots of sunshine but frequent drought. The wealth accrued and squandered from mining, followed by poverty and rampant unemployment, have brought the country to near collapse. Nauru's isolation and lack of attractions have also kept the tourists away: the island's most prominent imports are the hundreds of asylum seekers transported here by the Australian government in exchange for financial aid.

Nepal

Although hemmed in by giants – India on one side and China on the other – tiny Nepal still moves to its own unique rhythms. In this one-time Himalayan kingdom, Hinduism and Buddhism flow together like the rivers that gush down from the high Himalaya and drain out into the Indian plains. Tiered temples and medieval monasteries perch atop mountain passes and mountaineers and trekkers throng to trails that climb through rhododendron forests and high-altitude deserts to the summits of the world's tallest peaks. Then there are the overlooked lowlands, sprinkled with wilderness reserves that shelter tigers, elephants and rhinos. Disasters – both natural and man-made – have often pushed the country to the brink, but somehow, Nepal always bounces back. If there's a better stand-in for Shangri-La anywhere on Earth, we've yet to find it.

Best time to visit
September to November and March to May, to avoid summer rains and icy winters

Top things to see
- The tangled, temple-strewn backstreets of Kathmandu
- Stupendous stupas at Bodhnath and Swayambhunath
- The royal squares in Patan and Bhaktapur, earthquake scarred, but still magnificent
- Chariot parades and rainbow-coloured festivals in the Kathmandu Valley
- Views of the highest mountains in the world from viewpoints the length of Nepal

Top things to do
- Pit muscle against mountain on the trek to Everest Base Camp
- Buy singing bowls and Tibetan rugs in the bazaars of Kathmandu
- Drift though the medieval villages of the Kathmandu Valley, surrounded by temples and traders
- Raft or kayak on the wild white waters of the Bhote Kosi or Sun Kosi
- Seek rhinos and tigers amid the elephant grass in Bardia or Chitwan National Park

Getting under the skin
Read: *Arresting God in Kathmandu* by Samrat Upadhyay; or WE Bowman's mountaineering spoof *The Ascent of Rum Doodle*
Listen: to the evocative Nepali folk music of Sur Sudha
Watch: Eric Valli's classic *Himalaya*, with all characters played by Dolpo villagers; or *Even When I Fall* by Sky Neal and Kate McLarnon, documenting the lives of former trafficking victims who established Nepal's first circus
Eat: *dal bhat* (lentils, vegetables and rice) – served twice a day, every day if you go trekking
Drink: salted butter tea; *chang* (milky beer made from rice or barley); or hot *tongba* (millet beer)

In a word
Ke garne? (What to do?)

Celebrate this
Jumping ahead of most of the developed world, Nepal became one of the first countries to legally recognise non-binary gender, giving members of the *meti* (transsexual) community the right to identify legally as 'other' in 2007.

Random fact
Sagarmatha (Mt Everest) moved three centimetres southwest during the April 2015 earthquake.

1. The first stupa at Bodhnath was built sometime after 600 CE, when the Tibetan king, Songtsen Gampo, converted to Buddhism

2. Mountaineers descend from the summit of Island Peak in Khumbu Valley

3. Yaks, here from Namche Bazaar in Khumbu, are often used by sherpas to carry supplies into rugged regions

4. Lukla airport, the hub for treks to the Everest region, is regarded as one of the world's most terrifying

1. Watch windmills in action at the preserved village of Zaanse Schans

2. Cuypers Library in the Rijksmuseum is the Netherland's largest and oldest library of art history

3. Looking along a canal towards the Basilica of Saint Nicholas in Amsterdam

①

CAPITAL AMSTERDAM (GOVERNMENT SEAT IS THE HAGUE) // **POPULATION** 17.3 MILLION // **AREA** 41,543 SQ KM // **OFFICIAL LANGUAGE** DUTCH

Netherlands

The Netherlands is where the modern and the traditional fuse seamlessly. It's a nation where creativity flourishes, from innovative start-ups and cutting-edge architectural statements to eco-initiatives and a burgeoning food scene. This goes hand-in-hand with 17th-century canals, tulip fields, windmills, ancient churches, candlelit brown cafes and the masterpieces of Van Gogh and Rembrandt. Beautiful old cities such as Amsterdam, Haarlem, Leiden, Delft, Utrecht and many more attest to the wealth that fuelled the growth of the Netherlands – the ever-tolerant Dutch have always excelled at making the most of what they have. From the swamps and shallows of a land that often sits below sea level they've created a modern, comfortable country and one of the world's easiest places for cycling.

Best time to visit

April for tulips, May to October for cafe-friendly weather

Top things to see

- Amsterdam, one of Europe's best-preserved great cities, with canals and 17th-century architecture
- The striking, cutting-edge landmarks of one of Europe's most innovative cities, Rotterdam
- Dutch masterpieces by the likes of Rembrandt, Vermeer, Van Gogh and Mondrian at Amsterdam's Rijksmuseum and Stedelijk Museum
- Millions of springtime tulips at Keukenhof

Top things to do

- Bicycle along the flattest and best-maintained cycle paths in the world, such as the North Sea Cycle Route, which follows the Dutch coastline
- Retrace the life of Vermeer in tidy, pretty Delft
- While away an afternoon tucked cosily into one of the country's historic brown cafes

Getting under the skin

Read: the classic *Diary of Anne Frank*, which never loses its impact
Listen: to Tiësto, one of many internationally famous Dutch DJs
Watch: Paul Verhoeven's *Turks Fruit*, made before he found fame in Hollywood
Eat: *frites* (fries) with any of dozens of sauces; pancakes, served savoury or sweet
Drink: *jenever*, Dutch gin, made here for 400 years

In a word

Dag (Hello/goodbye)

Celebrate this

The ongoing need to keep water at bay led to the Delta Works project, which helped alleviate flooding with dams, sluices and barriers; rising sea levels mean similar schemes will be needed in coming decades.

Random fact

Around one-third of the Netherlands is below sea level, with the country's highest point, Vaalserberg hill, reaching a modest 322m.

CAPITAL WELLINGTON // POPULATION 5 MILLION // AREA 267,710 SQ KM //
OFFICIAL LANGUAGES ENGLISH, MĀORI & NEW ZEALAND SIGN LANGUAGE

New Zealand

As the planet heats up environmentally and politically, it's good to know that New Zealand exists. This uncrowded, green, peaceful and accepting country is the ultimate escape. Mother Nature decided to take her best features and condense them all in this South Pacific gem, from sublime forests, pristine beaches and snowcapped mountains to winding fjords and active volcanoes. The decadent scenery is not just there for admiring – it's for trekking through, paragliding over, hiking up or skiing down. Once you've exhausted yourself, take sleepy inland roads to find superb wineries and a thriving Māori culture before discovering the nation's cosmopolitan cities. With all this at their feet Kiwis have a distinct lust for life, and it's easy to understand why all who visit fall in love with their country.

Best time to visit
November to April for fun in the sun, June to August for fun in the snow

Top things to see
- Volcanic mud bubbles, spurting geysers and neon geothermal pools at Rotorua
- Rare kiwi birds in the wild on Stewart Island's remote beaches
- The icy grandeur of the Franz and Fox Glaciers from the ground or the air
- The multi-level maze of caves, canyons and underground rivers of the Waitomo region
- The ulimate meeting of forest, land and sea around Milford and Doubtful Sounds in Fiordland

Top things to do
- Kayak or hike the golden-sand beaches and turquoise inlets of Abel Tasman National Park
- Explore the lake and mountain scenery around Wanaka before adjourning to craft beer taprooms
- Negotiate the Tongariro and Poukaki Crossings – two of the planet's best one-day hikes
- Experience urban distilleries, cafes and restaurants in raffish and cosmopolitan central Wellington
- Discover diverse marine wildlife including whales, seals and pelagic birds around the rugged South Island coastline off Kaikōura

Getting under the skin
Read: *The Luminaries* by Eleanor Catton, a Booker Prize–winning novel set in 1860s New Zealand
Listen: to anything on the *Great New Zealand Songbook*, a compilation of the greatest Kiwi tunes from the last two centuries
Watch: *Hunt for the Wilder People*, from Kiwi cinema legends, Taika Waititi and Sam Neill
Eat: Innovative Māori-influenced cuisine from chef Monique Fiso at Hiakai restaurant in Wellington
Drink: craft beer made with Riwaka and Nelson sauvin hops from New Zealand's Tasman region

In a word
Kia ora (Hello)

Celebrate this
Following emphatic and effective responses to terrorism, natural disaster and Covid-19 throughout 2019 and 2020, New Zealand Prime Minister Jacinda Ardern earned kudos as one of the world's most admired national leaders.

Random fact
No matter where you are in New Zealand you are never more than 120km from the sea.

1. A traffic jam near Milford Sound on the South Island; there are more than 26 million sheep in the country

2. Beautiful Lake Tekapo and the Southern Alps

3. The mountain-dwelling kea parrot hangs out at trailheads and parking lots to rob cars of rubber seals

4. Modestly proportioned Wellington is New Zealand's capital

Best time to visit
November to March, to avoid both the worst of the rains and the dusty end of the dry season

Top things to see
- The moon rise over the cathedral in colonial Granada
- One island, two volcanoes on the biggest lake in Central America: Isla de Ometepe
- Caimans, howler monkeys and a Spanish fortress along the Río San Juan
- Bright handicrafts piled high at Masaya's Mercado Viejo
- Wooded hills and wildlife in the community-managed Miraflor Reserve

Top things to do
- Ride the waves in the laid-back surfer's haven of San Juan del Sur
- Climb up Cerro Negro then sandboard down its ash-covered slopes
- Dive into an aquatic world of hammerheads and eagle rays off Little Corn Island
- Sip coffee and talk politics with the intelligentsia in the charming university town of León
- Play castaway on the white sand of the Pearl Keys

Getting under the skin
Read: *Stories and Poems by Rubén Darío*, the father of Spanish modernism; Salman Rushdie's *The Jaguar Smile: A Nicaraguan Journey*; and Gioconda Belli's *The Country under My Skin: A Memoir of Love and War*
Listen: to electro-pop goddess Clara Grun; the smooth salsa of Luis Enrique; the Manu Chau–influenced Perrozompopo; and legendary marimba-folk artists Los Mejía Godoy
Watch: *Nicaragua Was Our Home* – a documentary about Miskito Indian repression; or the Oscar-nominated *Alsino and the Condor*, about a boy in war-torn Nicaragua
Eat: *nacatamales* (a mixture of cornmeal, potato, pork, tomato, onion and sweet chillies packed into a banana leaf and steamed to perfection)
Drink: Flor de Caña rum; *pinol*, toasted corn powder sweetened with sugar or taken with cacao (chocolate)

In a word
¡Va pue'! (All right!)

Celebrate this
Lake Nicaragua is one of the few lakes in the world to contain bull sharks (which hop up rapids to get here) and is the country's main source of water – but it's threatened by water loss and a proposed canal.

Random fact
Nicaragua is the largest country in Central America, as well as the least densely populated.

1. Looking over the San Juan River from Fortaleza fortress

2. First built in 1525, Our Lady of the Assumption Cathedral stands in central Granada

3. Relaxation Caribbean-style on Nicaragua's Little Corn Island

PHILIP LEE HARVEY | LONELY PLANET (ALL IMAGES)

N CAPITAL MANAGUA // POPULATION 6.2 MILLION // AREA 130,370 SQ KM // OFFICIAL LANGUAGE SPANISH

Nicaragua

Poet Rubén Darío called it 'Our America, trembling with hurricanes, trembling with love'. The last few decades have seen Nicaragua shaken by war and economic woes, and more recently widespread demonstrations have taken aim at an increasingly autocratic government. But Nicas, as they're known, are persistent and proud and their nation is an extraordinary place, home to colonial cities, big beach breaks, forested highlands and mystical islands. Nicaragua stretches from the towns and cities of the Pacific coast past lakes and volcanoes to some of Central America's most untouched rainforest, before landing on a Caribbean coast of Creole accents and dugout canoes. Dollars stretch far here, and a warm welcome eases the transition. Just don't offend your hosts by calling it 'the next Costa Rica'.

N | CAPITAL NIAMEY // POPULATION 22.8 MILLION // AREA 1,267,000 SQ KM // OFFICIAL LANGUAGE FRENCH

Niger

Home to two of the Sahara's most truly astounding sights – the Aïr Mountains and the great dune sea of the Ténéré – Niger is falling deeper and deeper into the desert's grasp. The march of the Sahara south is reducing arable and pastoral lands for a population that desperately needs it, and the rise of Islamic insurgents in the desert after the fall of Libya has led to more attacks across the country. Add regional insecurity due to Boko Haram, and Niger is no longer safe for travel. There can only be hope for Nigeriens that peace and some prosperity return to the Tuareg region of the north, the captivating Saharan towns such as Agadez, and to the great cities and cultures of the diverse south.

Best time to visit
October to February, when it's dry and relatively cool

Places of interest
- The Ténéré Desert, the Sahara's most spectacular and remote dune sea
- The remarkable Aïr Mountains, a barren and beautiful desert massif
- Agadez, perhaps the most romantic trans-Saharan caravan town, with its iconic pyramid-shaped mud mosque
- Zinder's ancient Hausa sultanate with a palace, weekly market and fine old quarter
- The Niger River town of Ayorou, which has a wonderful Sunday market and hippo-filled waters
- Kouré, home to the Sahel's last, highly endangered giraffe herd
- Parc Regional du W, where bush elephants and lions still roam

Local customs
- During La Cure Salée festival, Wodaabé men don elaborate make-up and dance in the hopes of being proposed to by a woman
- Buses, bush taxis and the country in general grind to a halt at *salat* (Muslim prayer)
- Traditional wrestling is wildly popular, and incorporates prayer, poetry and the wearing of *grigri* (charms)
- Due to the harsh dictates of the Sahara, some Tuareg forgo fasting during Ramadan

Getting under the skin
Read: *Riding the Demon: On the Road in West Africa* by Peter Chilson, a view of Niger via its bush taxis
Listen: to Mamar Kassey (*Alatouni*), or to Etran Finatawa (*Desert Crossroads*), Tuareg and Wodaabé bands famous for their desert blues
Watch: *The Sheltering Sky*, directed by Bernardo Bertolucci and filmed partly in Agadez
Eat: dates, yoghurt, mutton, rice with sauce and couscous
Drink: Bière Niger, a local lager

In a word
Na gode (Thank you)

Celebrate this
Ali Malam and Sakina Mati, local farmers who have managed to regenerate their desertified farmland and double their crop yields using simple, but novel techniques: they have gone on the radio to share their story to aid other farmers do the same.

Random fact
Deep in the Ténéré is a monument visible from space; the life-sized silhouette of a DC-10 aircraft sits within a dark stone compass surrounded by 170 broken mirrors, one for each person killed when UTA flight 772 (en route to Paris from Brazzaville) exploded in the sky here in 1989.

1. Ténéré is a very sparsely populated Saharan region bounded by the Aïr Mountains, the Hoggar Mountains, the Djado Plateau, the Tibesti Mountains, and the basin of Lake Chad
2. Look out for rock carvings of cattle and humans by ancient hunter-gatherers in Agadez
3. Tuareg women pump water at Timia Oasis in the Aïr Mountains
4. The salt mines in Niger's Sahara stand where once a vast sea did

CAPITAL ABUJA // **POPULATION** 214 MILLION // **AREA** 923,768 SQ KM // **OFFICIAL LANGUAGE** ENGLISH

Nigeria

A pulsating and populous powerhouse, Nigeria is West Africa's most dynamic country. And nowhere is this more evident than in its megacity, Lagos. Fuelled by oil money, and increasingly by tech and telecommunications, the former capital provides a sensory overload, where exploding art, music and cuisine scenes collide with traffic chaos, noise, pollution and crime. Outside the city it's a different world, proceeding at a different pace, from Yoruba shrines and languid river deltas to ancient trading cities and poignant slave ports. Sadly, the tension and violence associated with the growth of Boko Haram, which has made much of the Muslim northeast inaccessible to travellers, has splintered the country, and Nigeria's government is working to avoid it fracturing along religious and ethnic lines.

Best time to visit
November to February

Top things to see
- Intimidating and exciting in equal measure, Lagos, the capital in all but name
- Benin City, the rich legacy of ancient Benin with bronze casting, museum and the Oba's Palace
- The modern, made-to-measure Nigerian capital of Abuja, where amenities abound
- The historic old river port and primate conservation centres in the capital of Cross River state, Calabar
- Zuma Rock, a mammoth granodiorite monolith that towers 750m above the plains outside Abuja

Top things to do
- Climb Olumo Rock in Abeokuta to take in shrines, sacred trees, war-time hideouts and incredible vistas
- Wander in the Unesco–listed Osun Sacred Grove in Oshogbo, the centre for Yoruba spirituality
- Walk in the rainforest canopy at the Afi Mountain Drill Ranch, home to rehabilitated drill monkeys
- Make a musical pilgrimage to the New Afrika Shrine, the epicentre of Afrobeat in Lagos

Getting under the skin
Read: anything by Chinua Achebe, Ben Okri, Chimamanda Ngozi Adichie or Nobel Prize–winner Wole Soyinka; *Looking for Transwonderland: Travels in Nigeria* by Noo Saro-Wiwa for a powerful read
Listen: to Afrobeat, the Nigerian musical style mastered by the late Fela Kuti and his son and heir-apparent Femi Kuti; Burna Boy, Olamide, Rema and Mayorkun are other names to listen out for
Watch: *Half of the Yellow Sun*, based on Chimamanda Ngozi Adichie's novel about the Nigerian civil war; *Black November* by Jeta Amata, a hard-hitting look at Western exploitation of Nigeria's oil fields
Eat: 'chop' (food), such as pepper soup and *suya* (spiced kebabs)
Drink: Guinness Foreign Extra Stout

In a word
Na So (It is so)

Celebrate this
Jeta Amata, one of West Africa's most acclaimed filmmakers: he directed his first film at the age of 21.

Random fact
Almost one out of every five Africans is a Nigerian; in just two generations Lagos' population has grown from 200,000 to over 20 million.

1. The National Arts Theatre building in Lagos was designed to look like a military cap

2. Femi Kuti, son of Afrobeat legend Fela Kuti, performing; Fela Kuti's daughter, Yeni Anikulapo-Kuti, founded Nigeria's Felabration festival

3. A drill monkey in Calabar; the more colourful the male's rump, the higher his social standing

4. The Abuja National Mosque is open to non-Muslims

Best time to visit
April, for the national day celebrations, and September to October when humidity lessens

Top things to see
- Kilometres of landmines in the Demilitarized Zone (DMZ), dividing North and South Korea
- Serene mountain scenery at the hill resort of Kumgangsan
- The talent of young martial artists, musicians and gymnasts at Mangyongdae Children's Palace
- Sacred Paekdusan, mountain birthplace of Hwanung, the founder of the first Korean kingdom
- Pyongyang's Juche Tower, a three-dimensional embodiment of the Korean principle of self-reliance

Top things to do
- Meet locals at their most relaxed in the popular picnic spot of Moran Hill
- Step back into Korea's history at Kaesong, Korea's ancient capital
- Pay your respects at the Mansudae Grand Monument (but then, you have to)
- View Supreme Leaders Kim Il-sung and Kim Jong-il lying in state at the Kumsusan Memorial Palace of the Sun
- Soak up the surreal spectacle of thousands of perfectly choreographed people at the Arirang Mass Games

Getting under the skin
Read: *Nothing to Envy: Ordinary Lives in North Korea* by Barbara Demick; or the Pulitzer-winning *The Orphan Master's Son* by Adam Johnson
Listen: to the rousing patriotic anthems played on the Pyongyang metro
Watch: Daniel Gordon's *A State of Mind*, exploring the lives of North Korean gymnasts; or Ri Yun Ho's *The Story of Our Home*, a rare glimpse of North Korean filmmaking
Eat: *naengmyeon* (cold kudzu-flour or buckwheat noodles)
Drink: *soju* (local vodka); or Taedonggang, the national beer of North Korea

In a word
Juche (The national policy of 'Self Reliance')

Celebrate This
After defecting to the south in 2016, Thae Yong-ho – formerly North Korea's deputy ambassador to the UK – went on to become a leading figure in South Korean politics as National Assembly member for the Gangnam district of Seoul

Random fact
Now-deceased former Supreme Leader, Kim Jong-il, uttered only six words in public: 'Glory to the heroic soldiers of the People's Army!'

1. Construction began on Ryugyong hotel in 1987 and it remains the world's tallest unoccupied building

2. Looking over Pyongyang

3. Synchronised performances are a speciality of the Arirang Mass Games

N · **CAPITAL** PYONGYANG // **POPULATION** 25.6 MILLION // **AREA** 120,538 SQ KM // **OFFICIAL LANGUAGE** KOREAN

North Korea

Secretive only just describes North Korea, the insular northern half of the Korean peninsula. The Democratic People's Republic cut itself off from the outside world at the end of the Korean War, and now admits visitors only on strictly regimented tours. When not rattling sabres at South Korea, the nation spends much of its time on extravagant propaganda exercises, making for some spectacular, if surreal, tourist experiences. With an official escort, you can visit pristine, empty state institutions, empty skyscraper cities, empty mountain resorts and historic sites, and pay your respects to a massive statue of the supreme leader (this last one is usually obligatory). Only 5000 tourists peer over the wall ever year, and even fewer roam beyond the capital, Pyongyang, into Asia's least-explored frontier.

1. The dome of Treskavec Monastery features Byzantine frescoes

2. Alexander the Great surveys Macedonia Square in Skopje

3. The Church of Sveti Jovan at Kaneo is possibly North Macedonia's most photographed structure

N · **CAPITAL** SKOPJE // **POPULATION** 2.1 MILLION // **AREA** 27,713 SQ KM // **OFFICIAL LANGUAGES** MACEDONIAN & ALBANIAN

North Macedonia

North Macedonia is a fascinating combination of ethnicities, history and cultures that coalesce in a small, landlocked but immensely picturesque nation at the heart of the Balkans. With a Slavic Orthodox majority and thriving Albanian, Serbian and Turkish minorities, plus tiny pockets of Vlachs and Roma, this multicultural republic is on the up, having finally resolved the long-standing name dispute with neighbouring Greece. The former government worked hard at creating what they saw as a much-needed national consciousness, largely by peppering the capital city, Skopje, with monuments to historic Macedonians. Outside the capital mountain peaks soar, three-million-year-old Lake Ohrid glistens and a slow-food scene is blossoming in rural communities.

Best time to visit
April to September

Top things to see
- Ohrid's graceful domes and the diminutive 13th-century Church of Sveti Jovan at Kaneo
- Skopje's controversial collection of quirky statues, bridges and neoclassical-style buildings
- Marvel of Yugoslav architecture, the Ilinden Uprising Monument (aka Makedonium) in Kruševo
- Frescoes and Roman ruins at Treskavec Monastery
- The sublime Ottoman floral motifs of the Painted Mosque in Tetovo

Top things to do
- Haggle for traditional crafts, carpets and dolls in Skopje's Čaršija, the city's central Turkish bazaar
- Taste *vranec*, the signature full-bodied red, at Popova Kula winery in the Tikveš wine region
- Go hiking, mountain biking, horse riding or skiing in Mavrovo National Park

Getting under the skin
Read: *To the Lake* by Bulgarian writer Kapka Kassabova, set in Lakes Ohrid and Prespa
Listen: to anything by Toše Proeski, or Esma Redžepova, the Romani singer and humanitarian
Watch: *Before the Rain*, directed by Milcho Manchevski; or Oscar-nominated *Honeyland*, about one of Europe's last wild beekeepers
Eat: *skara* (grilled meat), accompanied by *šopska salata* (tomatoes, peppers, onions and cucumbers)
Drink: *rakija* (grape brandy) or hearty red wines from the Kavadarci region

In a word
Haydemo (Let's go)

Celebrate this
Restaurant owner Tefik Tefikovski helped launch the country's slow-food movement, promoting traditional cooking and sustainable farming.

Random fact
In 2008, 200,000 Macedonians planted two million trees in the first Macedonian Tree Day.

Best time to visit
April is typically the driest month; May to September promises the warmest temperatures

Top things to see
- Titanic Belfast, a multimedia visitor attraction situated where the famous liner was built
- The 40,000 hexagonal basalt columns of the Giant's Causeway, a Unesco-listed wonder
- Downpatrick, where St Patrick's mission began and ended, and home to a mighty cathedral
- The historic walls encircling the history-steeped city of Derry
- Strangford Lough's Castle Ward Estate, an iconic *Game of Thrones* filming location

Top things to do
- Wander among reconstructed farmhouses, forges and mills at the Ulster Folk Museum near Belfast
- Wobble across the narrow, swaying Carrick-a-Rede Rope Bridge on the Causeway Coast
- Surf the Atlantic breakers at the sweeping beaches around Portrush
- Learn the secrets of whiskey making on a behind-the-scenes tour of Old Bushmills Distillery
- Tee off at revered Royal County Down Golf Club in Newcastle

Getting under the skin
Read: *Eureka Street* by Robert McLiam Wilson, set before and after the 1994 ceasefires; *The Eggman's Apprentice* by Maurice Leitch, depicting contemporary rural life
Listen: to Van Morrison's wonderfully evocative music about his formative childhood haunts in and around Belfast
Watch: Oscar-winning short *The Shore* for insights into the Troubles; *Derry Girls* for a comedic, poignant teenage look at the momentous 1990s
Eat: bountiful Ulster fry breakfasts incorporating crispy golden-brown soda bread and potato farls
Drink: Bushmills whiskey or flavour-packed craft beers from 1981-established Hilden Brewery

In a word
Bout ye? (How are you?)

Celebrate this
George Best was a Northern Irish footballing legend, loved by fans for his skill – and the tabloids for his celebrity lifestyle. Playing mostly for Manchester United, he also represented his country 37 times and his legacy includes an eponymous local airport.

Random fact
At 392 sq m, Lough Neagh is Britain and Ireland's largest freshwater lake, big enough to swallow the city of Birmingham (UK or USA – either would fit).

1. The columns of the Giant's Causeway were formed when lava cooled and cracked as it reached the sea

2. The Dark Hedges of beech trees on Bregagh Road starred in *Game of Thrones*

3. Take a tour of Belfast's Titanic Quarter

CAPITAL BELFAST // **POPULATION** 1.9 MILLION // **AREA** 14,130 SQ KM // **OFFICIAL LANGUAGE** ENGLISH

Northern Ireland

Otherworldly geological formations, spectacular surf coast, magical glens and ancient standing stones make the northeastern corner of the Emerald Isle a fascinating place to explore, as does its complex history. Northern Ireland has been a separate country since the partition of Ireland in 1921, when six Ulster counties opted out of the Irish Free State to remain within the United Kingdom. The arising conflicts infamously peaked during the Troubles, but since the 1998 Good Friday agreement Northern Ireland has forged an inspiring path to peace. It has also reinvented itself as a thriving visitor destination, not only for its history but also its burgeoning drinking, dining and nightlife scenes and photogenic landscapes.

CAPITAL OSLO // POPULATION 5.5 MILLION // AREA 323,802 SQ KM // OFFICIAL LANGUAGE NORWEGIAN

Norway

Norway is the supermodel of Scandinavia, a peak- and fjord-blessed country that gives its neighbours a serious case of mountain envy. There's a reason why artists, photographers and outdoor enthusiasts rave over this country: at almost every turn stunning wilderness lurks to overwhelm the senses. Much of Norway lies above the Arctic Circle, home to the midnight sun's ceaseless light in summer and the polar night's swirling Northern Lights in winter. Set among these natural phenomena is some of the world's most scenic hiking and skiing. There's a rugged frontier feel to much of the country, but this is still Scandinavia – design-driven bars and hotels are never too far away.

Best time to visit
May to September for sunshine, December to February for skiing and the Northern Lights

Top things to see
- The perfectly preserved mining cottages of Unesco World Heritage-listed Røros
- Lofoten Islands – mountain-islands dotted with fishing villages so postcard-perfect they look fake
- The jawdropping beauty of the Geirangerfjord, by boat or on foot
- Bryggen, the old medieval quarter of Bergen, with its long timber buildings housing museums, restaurants and shops
- Oslo's Vigeland Park, with its walkway lined with photogenic statues of screaming babies and entwined lovers

Top things to do
- Take a trip on a Hurtigruten coastal steamer, heading north from Bergen
- Stare spellbound out the windows on the spectacular, seven-hour Oslo–Bergen train route
- Spot polar bears in the Arctic archipelago of Svalbard, the definitive polar-adventure destination
- Take the ultimate selfie without falling off Preikestolen (Pulpit Rock), high above Lysefjord
- Hike among the high peaks and glaciers of the sublime Jotunheimen National Park

Getting under the skin
Read: *My Struggle* by Karl Ove Knausgård, a six-volume series of controversial autobiographical novels and Norway's unrivalled publishing phenomenon
Listen: to the *Peer Gynt* suite by beloved classical composer Grieg; the synthtastic '80s sounds of a-ha; the cool electro stylings of Röyksopp
Watch: *Max Manus* (Man of War), Norway's biggest budget blockbuster, recounting the true story of a resistance fighter during WWII
Eat: *laks* (smoked salmon); warm *moltebær syltetøy* (cloudberry jam) with ice cream
Drink: coffee; *aquavit* (or *akevitt*; a potent liquor distilled from potato)

In a word
Skal vi gå på ski? (Shall we go skiing?)

Celebrate this
The polar bears outside don't realise it, but the Svalbard Global Seed Vault is an internationally important project. A million seed samples are kept in controlled conditions in a mountainside cave, ready to come to the rescue in case of crises elsewhere.

Random fact
'Ski' is a Norwegian word, and thanks to rock carvings depicting hunters travelling on them, Norwegians make a credible claim to having invented the sport.

1. Fishing cabins (or *rorbu*) on Sakrisøy in Norway's Lofoten Islands

2. The great driving road of Trollstigen is part of the Norwegian Scenic Route to Geiranger

3. Krystad lies in Nordland, the north of Norway

4. The Oslo Opera House has a roof designed by architects Snøhetta to be walked upon

JUSTIN FOULKES | LONELY PLANET | JUSTIN FOULKES | LONELY PLANET // ANDREW MONTGOMERY | LONELY PLANET

Best time to visit
October to April, for the most comfortable weather, or catch Dhofar's *khareef* (monsoon) season between June and September to see this region at its greenest

Top things to see
- Wadi Ghul, the 'Grand Canyon of Arabia' from a perch on Jebel Shams
- The latticed-window courtyard at the heart of Jabreen Castle, once an important centre of learning for Islamic law, astrology and medicine
- The beautiful *khors* (rocky inlets) of the remote Musandam Peninsula from a *dhow* (traditional wooden fishing vessel)
- Vistas of the Batinah Plain from the ramparts of Nakhal Fort
- The ancient ruins of Al Baleed that belonged to the 12th-century trading port of Zafar

Top things to do
- Shop for Omani and Indian artefacts or traditional textiles in the Mutrah Corniche's souq
- Hike though Wadi Shab, the verdant gorge that feels like paradise
- Listen to a world-class performance within the Royal Opera House Muscat, a harmonious composition of marble, inlaid wood and arabesque design
- Drive the dunes of the Sharqiya Sands to encounter nomadic, desert-living Bedouin
- Witness the night-time drama of sea turtles nesting at Ras Al Jinz

Getting under the skin
Read: *Celestial Bodies* by Jokha Al-Harthi, a fictional story of rural village life in Oman
Listen: to Salid Rashid Suri, a 20th-century sawt singer and oud player known as the Singing Sailor
Watch: *Operation Oman*, which tells of Britain's secret war in Dhofar
Eat: *harees* – steamed wheat, boiled meat, lime, chilli and onions garnished with *ma owaal* (dried shark); *shuwa* (marinated meat cooked in an earth oven)
Drink: *kahwa* – strong Omani coffee flavoured with cardamom

In a word
Tasharrafna (Nice to meet you)

Celebrate this
Jokha Al-Harthi is the first Omani woman to have her work translated into English: her novel *Celestial Bodies* won the 2019 Man Booker International Prize.

Random fact
The coastal oasis of Sohar will forever be remembered from the *Arabian Nights* as the starting point for Sinbad's epic journeys.

1. Preparing a meal in the Bait al Safah living museum in Al Hamra

2. The dunes of Oman's Empty Quarter are part of the world's largest sand sea

3. Fishing at dawn from a dhow near the isolated village of Kumzar on the Musandam Peninsula

O **CAPITAL** MUSCAT // **POPULATION** 4.7 MILLION // **AREA** 309,500 SQ KM // **OFFICIAL LANGUAGE** ARABIC

Oman

In a region increasingly known for grandstanding and flashes of extreme wealth, whether artificial palm-shaped islands, opulent theme parks or glass towers scraping the heavens, Oman differs by being a place where heritage, society and nature are all on centre stage. There is a strong sense of identity here, one where pride in its past is matched by an understanding and confidence in a future that is also rooted in higher education. Evidence of Oman's stirring history is seen in the great sweep of Bedouin tradition and in the extraordinary forts and traditional architecture. And the remarkable stages for its times of old remain – the sculpted sands of the Empty Quarter, the jagged ramparts of mountain ranges and the wildly beautiful coastline.

P   **CAPITAL** ISLAMABAD // **POPULATION** 233.5 MILLION // **AREA** 796,095 SQ KM // **OFFICIAL LANGUAGES** URDU & ENGLISH

Pakistan

Always off the beaten track, Pakistan is one of the last great frontiers for travellers seeking white-knuckle adventure. However, thanks to political instability and a seemingly endless insurgency, this is not a destination for the faint-hearted. Indeed, some areas are off-limits entirely; others can only be accessed with an armed guard. Which is all a great shame as Pakistan offers a surfeit of riches: ruined cities, Mughal mosques, fabulous forts, fascinating tribal culture, and stunning Himalayan scenery, plus one of the world's greatest road journeys, the breathless Karakoram Hwy to China. Nevertheless, with some cautious planning, it is perfectly possible to travel to Pakistan and discover a captivating Islamic civilisation that straddles the divide between the modern age and Mughal times.

Best time to visit
November to April in the south, May to October in the north

Top things to see
- The Mughal splendour of Lahore Fort and the Badshahi Masjid
- A courtyard with room for 300,000 of the faithful at the Faisal Masjid in Islamabad
- One of Asia's earliest civilisations at Moenjodaro
- Sufi shrines and spectacular Mughal monuments in historic Multan
- Pantomime sabre-rattling at the Wagah–Attari border crossing between Pakistan and India

Top things to do
- Eat mutton *biryani* – the perfect quick lunch – in the bazaars of Karachi
- Sway along to haunting *qawwali* singing (Sufi devotional music) at the Data Darbar in Lahore
- Rattle along the bone-shaking Karakoram Hwy to Kashgar in China
- Marvel at the outrageous ornamentation of Pakistan's ornately decorated trucks and buses
- Trek through elemental landscapes in the Pakistan Himalaya and the Karakoram, Pamir and Hindu Kush mountain ranges

Getting under the skin
Read: Kamila Shamsie's *Burnt Shadows* or Mohsin Hamid's *The Reluctant Fundamentalist* to see the world from a Pakistani perspective
Listen: to the ballads of Nusrat Fateh Ali Khan, the most famous performer of *qawwali*
Watch: Shoaib Mansoor's *Khuda Kay Liye* (In the Name of God), following two musicians whose lives are touched by geopolitics; or Asim Abbasi's *Cake*, a comedy showing Pakistan's gentler side
Eat: chicken *karahi*, the national curry of Pakistan
Drink: fresh mango juice; chai (tea); or *badam* milk, flavoured with almonds

In a word
Insha'Allah (If God wills it)

Celebrate this
After being shot for the simple act of going to school, activist Malala Yousafzai went on to become the youngest ever Nobel laureate, winning the Peace Prize in 2014 for her campaigning work to bring education to women worldwide.

Random fact
The Sufi mystics of southern Pakistan follow an esoteric interpretation of Islam with a focus on music, dancing and smoking marijuana.

1. Spring arrives in Hunza, famous for its apricots, with a blaze of blossom

2. The Hunza valley lies in the autonomous Gilgit-Baltistan region in the northeast of Pakistan

3. Badshahi Mosque, a Lahore landmark, dates from the Mughal era

4. The Pakistan Monument and museum is set on the western Shakarparian Hills in Islamabad

Palau

Above and below water, Palau showcases the best of Micronesia. The diving here is world renowned for its vivid reefs, deep-blue holes, WWII wrecks, caves, tunnels, giant clams and more than 60 vertical drop-offs. Back on land, you can spot exotic birds, crocodiles slipping into the mangroves and orchids flourishing in shady corners. The archipelago is incredibly diverse, from coral atolls and tranquil specks with haunting WWII pasts, to Babeldaob, Micronesia's second largest island. With signs everywhere promoting feel-good vibes (such as WAVE: Welcome All Visitors Enthusiastically!), it's just as easy to fall in love with the people here as it is with their particularly blessed geography.

Best time to visit
It's driest from February to April, although it's warm year-round and can rain anytime

Top things to see
- The mushroom-shaped limestone islets of the Rock Islands on a scenic flight
- The abundance of marine life at Blue Corner, one of Palau's most magical dive sites
- The history and art of Palau on display at the Belau National Museum
- Ngardmau Waterfall, the highest falls in Micronesia
- Eerie Japanese WWII ruins on the tiny island of Peleliu

Top things to do
- Dive by trees of black coral, mammoth gorgonian fans, sharks and sea turtles around Peleliu
- Snorkel the alien-like, stingless-jellyfish world of Jellyfish Lake
- Take it really, really easy on the charming island of Angaur
- Get off-road on an all-wheel driving tour of Koror or Babeldaob
- Wend your way along the majestic Ngerdorch River on a boat cruise

Getting under the skin
Read: *Words of the Lagoon: Fishing and Marine Lore in the Palau District of Micronesia*, marine biologist RE Johannes' account of the knowledge of Palau's fishermen
Listen: to *Natural…*, the first album by popular Palauan band IN-X-ES
Watch: *The Last Reef: Cities Beneath the Sea*, a 3D cinematic spectacular that explores Palau's underwater world
Eat: Palau specialities such as taro-leaf soup
Drink: an ice-cold amber ale from Red Rooster, Palau's only craft-beer brewery

In a word
Omelengmes (The concept of politeness and respect)

Celebrate this
All incoming visitors are required to sign the Palau Pledge, a promise to act in an 'ecologically responsible' manner while in the country; it's the first of its type in the world.

Random fact
Palau still has a scattering of live WWII ammunition in the bush; if you're caught trying to remove them (and don't get blown up for your efforts), you'll incur a hefty fine.

1. The 340 islands of Palau gained sovereignty in 1994

2. Under the surface of the western Pacific live many wonderful creatures, such as the nautilus

3. Check out the 200-year-old murals on traditional huts

P

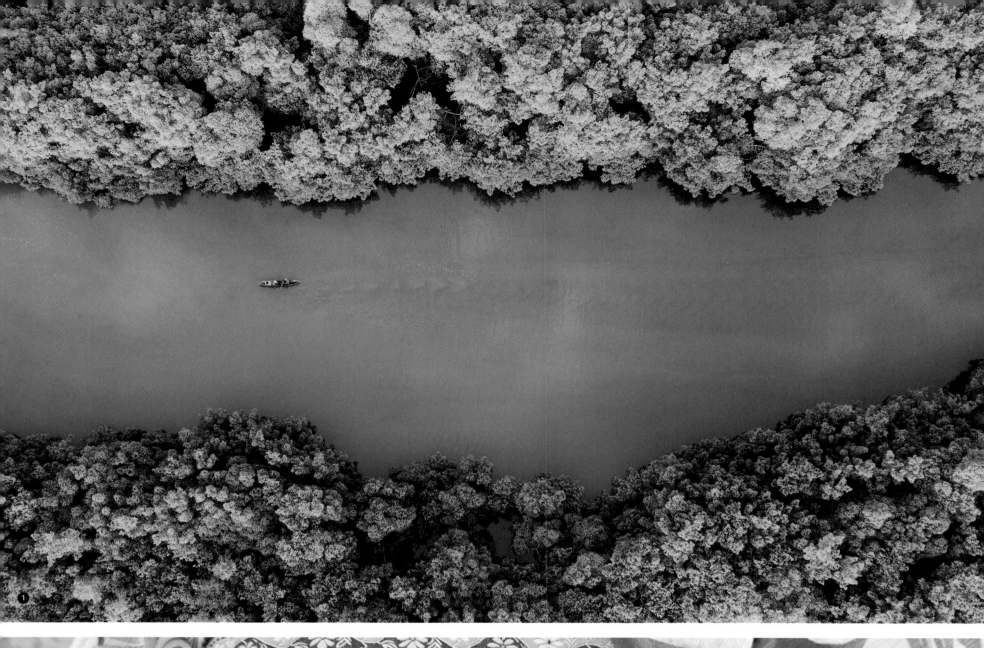

P **CAPITAL** PANAMA CITY // **POPULATION** 3.9 MILLION // **AREA** 75,420 SQ KM // **OFFICIAL LANGUAGE** SPANISH

Panama

Panama is the waistline of the Americas, cut in half by the world's greatest shortcut. The Panama Canal's 80km belt of locks brings the Atlantic to the Pacific, wedding east to west and helping define this small tropical nation. But Panama is a place to linger, not pass by. Pristine beaches, lush rainforest and big-city nightlife give a taste of the country's outstanding assets. Panama City is a major financial hub with a monumental skyline, yet one hour outside the city, indigenous Embera paddle dugout canoes and a couple of hours further, the roadless rainforests of the Darién harbour one of the world's greatest wildernesses. The Panama Canal expansion might signify increased business, but surf beaches, palm-strewn islands and coffee farms offer the chance to get away from it all.

Best time to visit
Mid-December to mid-April (the dry season)

Top things to see
- Monster freighters and capuchin monkeys sharing the Panama Canal
- The hip, regal and ruinous intersecting in the 17th-century Panama City neighbourhood of Casco Viejo
- Perfect beaches and poisonous dart frogs in the Bocas del Toro Archipelago
- The great trackless jungle of the Darién, where the primeval meets the present
- The serene 13km arc of sand at Playa las Lajas along the in-the-know surfers' paradise of the Gulf of Chiriqui

Top things to do
- Snorkel with sea turtles and sharks in Parque Nacional Coiba
- Trek through Parque Nacional Volcán Barú in search of the elusive quetzal
- Water-sopping and rum-soaked, revel on the crowded streets at Carnavales de Azuero
- Barter with the Guna on the sun-soaked autonomous San Blas islands
- Sip award-winning local coffee in the cool highland town of Boquete

Getting under the skin
Read: *The Path Between the Seas* by David McCollough, on the elephantine undertaking of the Panama Canal
Listen: to the salsa of Rubén Blades; the folk of Samy and Sandra Sandoval; Sech's autotuned reggaeton
Watch: the Academy Award–winning *The Panama Deception*, by Barbara Trent, which investigates the US invasion of Panama; *Rubén Blades is Not My Name*, an intimate portrait of the musical great
Eat: *sancocho* (chicken, yam and coriander soup); or *carimañolas* (meat-filled yucca puffs)
Drink: *chicheme* (sweetcorn, cinnamon and vanilla in milk); or *seco* (sugar-cane liquor served with milk and ice)

In a word
¡Chuleta! (Wow!)

Celebrate this
The border between North and South America is set either at the Panama Canal, or the Darién Gap; from the summit of Volcán Barú, meanwhile, you can see the Atlantic and Pacific oceans at once.

Random fact
Panama hats originated in Ecuador – they acquired their name as they were distributed from Panama.

1. Boats are the best way to travel throught the Bocas del Toro islands on the Caribbean coast
2. Traditional Guna bracelets and *molas* (embroidered textiles) in the San Blas archipelago
3. Panama City is the base for a lot of banks and shipping companies, bringing in about 55% of Panama's GDP

Best time to visit
June to September is cooler, drier and takes in the majority of the provincial celebrations and Highlands *sing sings* (festivals or dances)

Top things to see
- The dancing, singing, whistle blowing and sometimes magic involved in a crowded game of Trobriand cricket
- A traditional *sing sing* celebrating the ascension of a chief, initiation rites and more
- The rumbling, billowing string of volcanoes in New Britain, which often puts on spectacular lava displays
- Astonishing biodiversity, including 38 species of birds of paradise, which put on fabulous displays in Varirata National Park and other forest areas

Top things to do
- Travel up the Sepik River to discover a treasure trove of quintessentially primitive Pacific art
- Trek into PNG's wild Highlands for glimpses into astonishing Indigenous culture and views from the lofty summit of Mt Wilhelm
- Dive shipwrecks, lost planes from WWII and reefs teeming with macro and megafauna, including whale sharks and mantas
- Follow in the muddy footsteps of WWII soldiers on the challenging, leech-infested Kokoda Track

Getting under the skin
Read: Tim Flannery's *Throwim Way Leg*, describing the author's trips in search of tree kangaroos; or Kira Salak's travelogue *Four Corners: A Journey into the Heart of Papua New Guinea*
Listen: to Telek's *Serious Tam*, showcasing the extraordinary voice of this native Papuan singer
Watch: Robin Anderson and Bob Connelly's 'Highlands Trilogy' documentary series – *First Contact, Joe Leahy's Neighbours* and *Black Harvest* – an outstanding exposition of Highlanders' first encounters with the outside world
Eat: *sasak* (sago) in the swampy Sepik; *kaukau* (sweet potatoes) in the Highlands; fresh fish and lobster on the coasts
Drink: PNG Highland-grown Arabica coffee

In a word
Em nau! (Fantastic! Right on!)

Celebrate this
One of just 17 megadiverse countries, PNG is home to almost 14,000 plant species, two-thirds of which are only found here, and scientists estimate a further 4000 local species still await discovery.

Random fact
More than 850 languages are spoken in PNG with adults speaking an average of three languages each.

1. Mt Hagen's Show is an important event for tribal performers

2. Papua New Guinea is home to 38 species of birds of paradise (including this cendrawasih) and unforgettable for bird watchers to visit

3. A Daga longhouse in the Southern Highlands

CAPITAL PORT MORESBY // POPULATION 7.2 MILLION // AREA 462,840 SQ KM //
OFFICIAL LANGUAGES ENGLISH, TOK PISIN & HIRI MOTU

Papua New Guinea

Less than 4km separate Papua New Guinea from the nearest point in
Australia, but the differences could not be more pronounced. Where
Australia is largely parched and dry, mountainous PNG is draped in a
steamy cloak of tropical rainforest. And where Australia tops global
rankings for quality of life, PNG languishes at the bottom of the list of
impoverished Asian nations. But there are more similarities than you
might expect, with the tribes of PNG sharing a rich history with Australia's
Indigenous people. The exotic birds and beasts that inhabit the dense
jungles, including tree-dwelling kangaroos, form a biodiverse bridge
between Australia and Asia. Most visitors skip the capital for the coast's
pristine coral reefs or the traditional culture of the interior.

Best time to visit
May to September (winter)

Top things to see
- The engineering behemoth of Itaipú, one of the world's largest hydroelectric dams and supplier of 90% of Paraguay's energy
- The haunting 18th-century colonial remnants of the Jesuit Missions of Trinidad and Jesús, both Unesco World Heritage sites
- Hummingbirds and toucans in verdant Parque Nacional San Rafael
- The colourful Carnaval of Encarnación, smaller but no less wild than Rio's big fest

Top things to do
- Go wildlife watching in the Mbaracayú Biosphere Reserve, one of the most biodiverse places on the planet
- Take off horseback riding, camping or simply admiring the beauty of Laguna Blanca
- Board a slow boat up the Río Paraguay to the wet and wild Pantanal
- Visit the Mennonite colonies before seeing birdlife and battle sites in the Chaco

Getting under the skin
Read: *I the Supreme* by Augusto Roa Bastos, a fascinatingly complex novel delving into the mind of 19th-century dictator 'Dr Francia'
Listen: to 'Pajaro Campana', a folk song based around the call of the bellbird (Paraguay's national bird); the Latin folk of Luis Alberto del Paraná
Watch: *The Mission* for depictions of Guaraní and Jesuit settlements in colonial days; *7 Boxes*, a taut thriller set in an Asunción market
Eat: succulent cuts of *tapa de cuadril* (similar to rump steak) and *chipas* (manioc, cheese and cornmeal rolls)
Drink: *tereré* (iced herbal tea); and *mosto* (sugarcane juice)

In a word
Mba'eichapa? (How are you? in Guaraní)

Celebrate this
Nine out of ten Parguayans speak the indigenous Guaraní or Jopará (a fusion with Spanish), while half the rural population only speak Guaraní. Almost everyone is at least part Guaraní (it's sometimes used as a synonym for Paraguayan), and it's a proud part of national identity.

Random fact
The War of the Triple Alliance (1864–70) devastated Paraguay, which lost a quarter of its territory and may have lost half of its population battling Argentina, Brazil and Uruguay.

1. A toucan in Paraguay; the birds use their oversized bills to cool down their bodies

2. The Jesuit missions of Trinidad were intended for the edification of the indigenous Guaraní

3. The jungle around the town of Colonia Independencia, which attracted European immigrants after WWII

P · **CAPITAL** ASUNCIÓN // **POPULATION** 7.2 MILLION // **AREA** 406,752 SQ KM // **OFFICIAL LANGUAGES** SPANISH & GUARANÍ

Paraguay

Dwarfed by Brazil and Argentina, small, landlocked Paraguay is sometimes described as South America's 'forgotten country'. Like its better-known neighbours, football madness and a burgeoning beef industry are national hallmarks. Paraguayans, however, have followed a different course through history. Most citizens, proudly touting their Guaraní heritage, scratch out a living at small-scale microenterprises or subsistence farming. The country is a remarkable study in contrasts, with horse-drawn carts sidling up to luxury automobiles, while huge Mennonite farms and rustic *campesino* (peasant farmer) settlements share space on the dusty, hard-scrabble Chaco. Paraguayans are famously laid-back, quick to share a *tereré* (iced herbal tea) with a visitor over long siestas in the sticky tropical heat.

Best time to visit
May to September (the dry season)

Top things to see
- Machu Picchu, the great Inca ruins hidden deep in mist-covered cloud forest
- Cuzco, a beautiful Andean town with Inca-made walls, cobblestone streets and gilded colonial churches
- The enchanting islands on Lake Titicaca, one of the world's highest navigable lakes
- Arequipa, a charming colonial city near smouldering volcanoes and the world's deepest canyons
- Parque Nacional Manu, home to anacondas, tapirs and great flocks of macaws

Top things to do
- Eat steaming street food or gourmet fusion dishes in food hotspot Lima
- Hike the Santa Cruz trail through the towering peaks of the Cordillera Blanca
- Charter a flight over the Nazca Lines, the mystical drawings in the earth left by a past civilisation over 1500 years ago
- Visit the ruins of Chan Chan, near Trujillo, once the largest pre-Columbian city in the Americas

Getting under the skin
Read: Mario Vargas Llosa's famed *Conversations in the Cathedral* about power and politics in 1950s Peru
Listen: to Susana Baca's Afro-Peruvian rhythms or the swinging pop-rock of Gian Marco
Watch: Claudia Llosa's award-winning fable *Madeinusa* about the clash between old and new in a somewhat surreal Andean town
Eat: *ceviche* (fresh seafood marinated in chilli and lime); *lomo saltado* (beef stir fry)
Drink: Inka Kola (a bubblegum-flavoured soft drink); *pisco* (a white grape brandy)

In a word
Buenos dias (Good day)

Celebrate this
Weaving has been part of Peruvian culture for 10,000 years; visitors can support this vibrant tradition by buying fabric and clothes from local cooperatives.

Random fact
Peru's pre-Columbian civilisations left such riches that treasures are still being unearthed – in 2019 a drone discovered a settlement older than Machu Picchu, 1500m above the famous site.

1. A traditional settlement on the banks of a Peruvian tributary of the Amazon River

2. The city of Trujillo has a 17th-century cathedral and pre-Columbian ruins nearby

3. Centuries ago, the Uros people constructed the floating islands on Lake Titicaca in order to escape more aggressive mainland ethnic groups, such as the Incas

4. Instead of Machu Picchu, consider visiting the remote Inca city of Choquequirao built above the Apurimac valley

P CAPITAL LIMA // POPULATION 31.9 MILLION // AREA 1,285,216 SQ KM // OFFICIAL LANGUAGES SPANISH, QUECHUA & AYMARA

Peru

The birthplace of the great Inca civilisation, Peru remains deeply connected to its heritage. Nine out of ten Peruvians have indigeous ancestry, and here Quechua-speaking highlanders mingle in markets beneath Andean peaks, while Uros peoples eke out a living on the floating islands of Lake Titicaca. A world away, remote tribes (some uncontacted) live deep in the Amazon. Peruvian cities provide a vivid contrast, a blend of frenetic and cosmopolitan neighbourhoods, scenic Spanish-colonial centres and buzzing plazas. The country's epic rainforests, coast and mountains are attracting more and more visitors while, despite regular political scandals, the economy is on the up. Innovative ideas – like sustainable tourism projects run by indigenous groups – are helping to ensure Peru's treasures will be around for generations to come.

P CAPITAL MANILA // POPULATION 109.1 MILLION // AREA 300,000 SQ KM // OFFICIAL LANGUAGE FILIPINO

Philippines

Scattered like emeralds across the Pacific Ocean, the 7000-plus islands of the Philippines are thatched by palm trees, garlanded by coral reefs, and overlooked by the vast majority of travellers to Southeast Asia. However, word is slowly getting out about the charms of Asia's largest Catholic country. Culturally, the islands are a surreal hotpot of American bravado, Chinese entrepreneurialism, tribal flamboyance and Spanish mysticism, set in a classic Southeast Asian landscape of coral beaches and rainforest-cloaked volcanoes. Sex tourism – a legacy of the Vietnam War – still casts a long shadow. But, increasingly, the Philippine archipelago is becoming known for white-sand beaches, frenetic fiestas and the spectacular dive sites and surf breaks that dot a staggering 36,000km of coastline.

Best time to visit
October to May, to avoid the worst of the typhoon season

Top things to see
- The almost-too-perfect sands of Boracay Island, with tourist numbers now carefully controlled
- Rice terraces on an epic scale at Banaue
- *Butanding* (whale sharks) on their annual migration past Luzon Island
- At least one fiesta; Filipino festivals are as flamboyant and energetic as the islands themselves
- The spooky hanging coffins of Sagada, a flashback to pre-Christian tribal death rites

Top things to do
- Ride in a jeepney – the wildly decorated stretched jeeps that serve as buses across the Philippines
- Dive into an eerie graveyard of WWII shipwrecks at Coron on Busuanga
- Survive a night out in Manila – there's a growing buzz to this once-notorious metropolis
- Meet the wildlife, be it graceful thresher sharks or tiny tarsiers, one of the world's teeniest primates
- Test your motorcycling skills on a dirt-road adventure in rugged Palawan

Getting under the skin
Read: *In Our Image* by Stanley Karnow, a harrowing exposé of the American colonial period; or F Sionil José's Spanish-era epic *Dusk*
Listen: to the sentimental croonings of Jose Mari Chan; or the agreeable Pinoy-rock of Eraserheads
Watch: Ishmael Bernal's emotional classic *Himala*; or Jerold Tarrog's *Goyo: The Boy General*, a patriotic biopic telling the story of Gregorio del Pilar, the youngest general in the revolutionary war
Eat: Spanish-influenced *adobo* (pork or chicken stewed in vinegar and soy sauce); or Chinese-inspired *pasit canton* (fried noodles)
Drink: Tanduay rum, typically served Cuba libre–style with Coke and a twist of lime

In a word
Susmaryosep – a vocal exclamation fusing Jesus, Mary and Joseph!

Celebrate this
Inspired by childhood experiences during the Japanese occupation, Reynaldo 'Nandy' Pacheco has devoted a lifetime to campaigning for a gunless society, challenging the state-supported culture of gun-ownership in the Philippines.

Random fact
Every Easter, devout Filipino Catholics offer themselves up to be temporarily crucified with real nails at San Fernando de Pampagna.

1. Pilgrims gather in Cebu City town square for the epic annual Sinulog Festival
2. The beautiful beach at Entalula Island off the coast of El Nido
3. Jeepneys are a popular form of public transport
4. The Sinulog Festival is the centre of the Feast of Santo Niño (the Christ Child) celebrations in Cebu

Best time to visit
May to September

Top things to see
- The architectural treasures of the Old Town and Wawel Hill in the former capital of Kraków
- The Warsaw Rising Museum, testament to the bravery and determination of local Poles
- Malbork Castle, the largest Gothic castle in Europe and once headquarters of the Teutonic Knights
- Port city Gdańsk with its monuments to Poland's Solidarity movement and white-sand beaches
- The Wieliczka salt mine, a subterranean labyrinth of salt-hewn chapels, statues and monuments

Top things to do
- Pay your respects at the infamous Nazi death camps of Auschwitz and Birkenau
- Spot a European bison emerging out of the undergrowth at Białowieża National Park
- Sail or canoe on the Great Masurian Lakes, connected by some 200km of canals
- Join the pilgrims trekking to see the Black Madonna at the Jasna Góra monastery
- Hike into the Tatra Mountains, the highest range within the Carpathians, or catch the cable car to Mt Kasprowy Wierch
- Ski between snow-covered pine trees in pretty Zakopane

Getting under the skin
Read: travel-themed novel *Flights* by Olga Tokarczuk, the 2018 winner of the Man Booker International Prize
Listen: to the works of Frederic Chopin; or to Henryk Gorecki's *Symphony No 3*
Watch: *Cold War* by Paweł Pawlikowski, a tragic love story in and out of Poland in the post-WWII period
Eat: *pierogi* (dumplings stuffed with minced meat); *oscypek*, salty sheep's cheese served with berry relish; *borscht*, either hot or cold, depending on the season; and *bigos*, a hearty hunter's stew of cabbage, sausage meat and juniper
Drink: vodka, especially flavoured varieties like cranberry and bisongrass; or Polish beers like Tyskie and Okocim

In a word
Dzien dobry (Good day)

Celebrate this
In recent years several Polish cities have embraced their Jewish heritage. Kraków has led the way in the revival of Jewish culture, especially in the former Jewish quarter of Kazimierz and through the annual Jewish Culture Festival.

Random fact
Nobel Prize–winning physicist Marie Curie completed her studies in Paris because, as a woman, she was denied a place at Kraków University.

1. Castle Square in the centre of Warsaw's Old Town

2. European bison have been reintroduced to the wild in Białowieża National Park

3. The lively city of Gdańsk sits beside the Baltic Sea

P · CAPITAL WARSAW // POPULATION 38.2 MILLION // AREA 312,685 SQ KM // OFFICIAL LANGUAGE POLISH

Poland

Stretching from the Baltic Sea to the Carpathian Mountains, Poland is a land whose fortunes have waxed and waned over centuries. Having shrugged off the Soviet mantle, it is now embracing modernity with energy and passion. With its upbeat capital, Warsaw, and the timeless elegance of Kraków, industrial heartland cities and open-air folk-architecture museums, urban Poland is nothing if not varied. Meanwhile, the Baltic resorts' white-sand beaches, water sports on the Great Masurian Lakes, the primeval forest of Białowieża National Park and the rocky gorges of the Tatra Mountains make the choice of outdoors thrills a difficult one. And these days, vegetarian food and artisan coffee are increasingly common alongside plates of dumplings and a glass of vodka.

P · CAPITAL LISBON // POPULATION 10.3 MILLION // AREA 92,090 SQ KM // OFFICIAL LANGUAGES PORTUGUESE & MIRANDESE

Portugal

Portugal's charms begin in its capital, Lisbon, where *miradouros* (viewpoints) perch like birds' nests on its seven hills and vintage trams screech past *azulejo*-tiled houses, old-school shops, retro bars and pearl-white Age of Discovery monuments, affording tantalising glimpses of the Rio Tejo. Venturing north brings you to the surf-lashed coast of Estremadura, Coimbra and its 500-year-old university, and the soulful city of Porto, with its grand parade of port lodges, strollable medieval centre and high-spirited nightlife. Further north, granite peaks and river valleys unfold, while the rural, food-focused Alentejo region and the cliff-backed beaches and whitewashed villages of the Algarve entice to the south. Out in the Atlantic, Portugal's lushly volcanic islands, Madeira and the Azores, are hiking heaven.

Best time to visit
March to June and September, July and August to bake on the Algarve's busy beaches

Top things to see
- The raw, windswept, end-of-the-world beauty of Cabo de São Vicente
- The uplifting and extraordinarily intricate Mosteiro dos Jerónimos in Lisbon's Belém quarter
- World-class modern art, Moorish architecture and fairy-tale palaces in Sintra
- Coimbra, home of Portugal's most celebrated university, with a medieval heart and live *fado* in student-packed bars
- Walled 14th-century Évora, a Unesco World Heritage Site

Top things to do
- Uncover cool boutiques and a heady mix of bars, restaurants and clubs in chilled-out Lisbon
- Poke around the Alfama quarter, Lisbon's castle-topped Moorish neighbourhood
- Explore dramatic cliffs, gold-sand beaches and scalloped bays on the Algarve
- Tour a port-wine lodge and taste Portugal's legendary tipple in popular Porto
- Hike rugged peaks in the Parque Nacional da Peneda-Gerês

Getting under the skin
Read: the funny 18th-century love story *Memorial do Convento* (Baltasar & Blimunda) by Nobel Prize-winner José Saramago
Listen: to Mariza, whose album *Terra* fuses traditional *fado* with world music
Watch: *Blood of my blood* by Portuguese director João Canijo, a view of a Lisbon life turned upside down
Eat: *cataplana* (seafood and rice stew in a copper pot); *pastéis de nata* (custard tarts)
Drink: Sogrape's Barca Velha *vinho* (wine) with a meal, followed by vintage port from the Douro Valley

In a word
Bom dia (Hello)

Celebrate this
Expressive, melancholy *fado* (fate) music has been part of the Portuguese identity for at least 200 years. Communicating a sense of *saudade* (longing) through local and Afro-Brazilian styles, it's on Unesco's list of Intangible Cultural Heritage.

Random fact
Bibliophiles should make a beeline for Lisbon's Livraria Bertrand, the world's oldest bookshop, keeping readers happy since 1732.

1. The seaside town of Ericeira is popular with surfers
2. The University of Coimbra is one of the oldest universities in Europe
3. Trams are the best way to explore the steep cobblestoned streets of Lisbon's Alfama district
4. A goat herder takes the high ground in the Parque Nacional da Peneda-Gerês

Q CAPITAL DOHA // POPULATION 2.4 MILLION // AREA 11,586 SQ KM // OFFICIAL LANGUAGE ARABIC

Qatar

Qatar's rulers seem determined to put the country firmly on the international map both as a forward-thinking financial and cultural centre and major regional political player. In recent years, Qatar's growing stature has put it at odds with its neighbours. The nation's *Al-Jazeera* TV network is an ongoing bone of contention. For travellers, Qatar's ambitious vision is plain to see in Doha, where massive investment has been channelled into creating some of the Gulf's most impressive heritage, culture and art developments. Although travel here remains mostly about the capital and its architectural showmanship, worthy city-escapes include the sand dunes around Khor Al-Adaid and the old coastal pearling villages along the way to Al-Zubarah Fort.

Best time to visit
October to April

Top things to see
- National Museum of Qatar, for a local-heritage-focused collection housed under a roof of multi-layered discs designed by Jean Nouvel
- Doha's Museum of Islamic Art, for the world's largest Islamic art collection
- The historic, redeveloped Souq Waqif to capture a sense of traditional architecture
- Sheikh Faisal bin Qassim Al-Thani Museum, for an eclectic ethnographic collection ranging from carpets to vintage cars
- The petroglyphs of Jebel Jassasiya

Top things to do
- Head off-road through desert dunes to the 'inland sea' of Khor Al-Adaid
- Have an up-close encounter with falcons at Doha's Falcon Souq
- Road-trip Qatar's northern tip to Al-Zubarah Fort, Qatar's only Unesco World Heritage Site
- Stroll the restored seafront souq of the old pearling village of Al-Wakrah
- Explore Doha's thriving contemporary art scene at Mathaf, Gallery Al-Riwaq and Katara Cultural Village

Getting under the skin
Read: *The Corsair* by Abdulaziz Al-Mahmoud; and *Jassim the Leader: Founder of Qatar* by Mohamed Althani
Listen: to Ali Abdel Sattar, Qatar's popular musical export
Watch: *Qatar: A Quest for Excellence*, exploring Qatari culture and music
Eat: *machboos* (heavily spiced meat and rice casserole; considered the national dish)
Drink: *karak* tea (cardamom-spiced milk tea made with evaporated milk)

In a word
Masha'Allah (God has willed it)

Celebrate this
With over 1.5km of galleries, the 2019-opening of the National Museum of Qatar finally gives the country a dedicated space for Qatari heritage to shine.

Random fact
Qatar is the second-flattest country in the world. Its highest point reaches a mere 203m.

1. Jean Nouvel's design for the National Museum of Qatar in Doha was inspired by the desert rose

2. The Doha city skyline viewed from the Museum of Islamic Art

3. Doha's atmospheric Souq Waqif was built on an ancient Bedouin trading site

R CAPITAL BUCHAREST // POPULATION 21.3 MILLION // AREA 238,391 SQ KM // OFFICIAL LANGUAGE ROMANIAN

Romania

Transylvania is hands-down Romania's best-known region, and Count Dracula its best-known resident. But Romania has a great deal more to it than vampire legends. With a mighty section of the Carpathian Mountains, a white-sand stretch of Black Sea coast, bucolic vistas wherever you turn your gaze, Orthodox churches and a clutch of medieval walled cities, Romania is more picturesque than Bram Stoker would have you believe. The country's more recent cultural draws include Cluj-Napoca's art collective, the Fabrica de Pensule; Iaşi's restored Palace of Culture; and acclaimed film festivals in Braşov and beyond. And from the protected wetlands of the Danube Delta to the well-marked trails in the soaring Retezat Mountains, Romania's great outdoors are teeming with adventure.

Best time to visit
May to June and September to October

Top things to see
- The wooden churches of Maramureş, with their intricate carvings and Gothic spires
- The world's second-biggest administrative building, Ceauşescu's imposing Palace of Parliament in Bucharest
- Dazzling icons and frescoes full of biblical scenes and allegories in Bucovina's painted monasteries
- The medieval citadel of Sighişoara, perched on a hillock and ringed with 14th-century towers
- The Merry Cemetery in Săpânţa, where colourful tombstones tell irreverent tales of the interred

Top things to do
- Uncover the medieval delights of Braşov, centre of bucolic Transylvania, home to Gothic churches and Europe's narrowest street
- Check over your shoulder as you tour so-called Dracula's Castle in Bran or venture to the real deal, Vlad the Impaler's citadel in Poenari
- Brave the white-knuckle, high-altitude Transfăgărăşan Road, celebrated by *Top Gear* as one of the world's most exciting drives
- Push a rowboat out into the Danube delta to tour an expansive wetland teeming with birdlife
- Soak up the sun on the golden sand of the most popular Black Sea resort, Mamaia

Getting under the skin
Read: *Diary of a Short-Sighted Adolescent* by Mircea Eliade, a schoolboy's perspective on life in early 20th-century Bucharest
Listen: to the inspirational, upbeat, improvised Romany mayhem of Taraf de Haidouks
Watch: Cristian Mungiu's *Four Months, Three Weeks and Two Days*, a drama about illegal abortions that won the Palme d'Or at Cannes and kicked off the Romanian New Wave
Eat: *mămăligă* (a cornmeal staple); *sarmale*, cabbage rolls crammed with mincemeat; or *ciorbă de burtă* (tripe soup that allegedly cures hangovers)
Drink: local wines, including Murfatlar, Odobeşti and Târnave; *ţuică*, a brandy potent enough to knock your socks off

In a word
Buna (Hello)

Celebrate this
As part of the Rewilding Europe project, magnificent European bison have been successfully reintroduced in Romania's southern Carpathian Mountains, taking the local population of the species up to 53 in 2018.

Random fact
In 1884, Timişoara became the first European city to have electric street lighting.

1. The lush green valleys and rolling hills of rural Romania

2. The opulence and extravagance of Bucharest's communist-era Palace of Parliament

3. The way of life of shepherds has remained unchanged for centuries and folk culture is alive and well

1. Ice stalactites in a cave at the shore of Lake Baikal

2. The neo-Byzantine ceiling of the 75m-high cupola of St Petersburg's Naval Cathedral

3. Russia's rail network is one of the largest in the world

R CAPITAL MOSCOW // POPULATION 141.7 MILLION // AREA 17,098,242 SQ KM // OFFICIAL LANGUAGE RUSSIAN

Russia

Stretching from the Baltic to the Bering Sea, Russia is a country of epic proportions. The original kingdom of Rus grew up along the western borders of Europe, where royal cities such as Moscow and St Petersburg still stun with their contemporary creativity. After the extravagance of the tsars came the austerity of the Soviet Union, which created its own set of imposing tourist attractions. Nevertheless, hints of Russia's golden age endure in lavish palaces, grand museums and some of the world's finest opera and ballet houses. That is one Russia; the other is a land of endless forests, silent lakes, active volcanoes, ancient fortresses, onion domes and gingerbread cottages. Come with an open mind and at all times be ready to toast Mother Russia with a glass of vodka.

Best time to visit
May to October

Top things to see
- Moscow's Kremlin, home to the treasures of the tsars and the institutions of the Russian state
- St Petersburg's Hermitage Museum with one of the world's finest art collections
- Kamchatka, the remote 'land of fire and ice' with its snow-covered volcanoes and reindeer herds
- The rural side of Russia, in bucolic country towns of the Golden Ring such as Suzdal

Top things to do
- Hit the beaches and ski slopes of the Black Sea coast or climb Europe's highest peak, Mt Elbrus
- Hike around ancient Lake Baikal or sail the Volga on a cruise from Kazan to Volgograd
- Take a 9289km journey on the Trans-Siberian Railway from Moscow to Vladivostok

Getting under the skin
Read: Boris Akunin's historical detective tale *Turkish Gambit*, Viktor Yerofeyev's erotic novel *Russian Beauty*, or Viktor Pelevin's sci-fi story *S.N.U.F.F.*
Listen: to Rachmaninov piano concertos, Tchaikovsky's lyricism and Stravinsky's modernism
Watch: *Leviathan* by Andrei Zvyagintsev, a tale of one man's struggle against official corruption in northern Russia
Eat: *pelmeni* (dumplings stuffed with meat), or borsch with *smetana* (sour cream)
Drink: vodka, Baltika beer and piping hot tea poured from a *samovar*

In a word
Zdrastvuyte (Hello)

Celebrate this
A 2015 survey by the World Wildlife Fund found as many as 540 Amur tigers living in the Russian Far East, a significant increase over previous figures.

Random fact
The Hermitage in St Petersburg is home to a crew of more than 70 cats, who keep the palace rodent-free as they have since the time of Catherine the Great.

1. Close encounters with silverback mountain gorillas can be had in Volcanoes National Park

2. Mt Muhabura rises over a Rwandan farming landscape

3. Intore dancers perform at Kwita Izina, the annual gorilla naming ceremony

CAPITAL KIGALI // POPULATION 12.7 MILLION // AREA 26,338 SQ KM // OFFICIAL LANGUAGES KINYARWANDA, ENGLISH & FRENCH

Rwanda

Rwanda, 'The Land of a Thousand Hills', is draped with life: mountain gorillas play in pockets of virgin rainforest on volcanoes, newly reintroduced lions and rhinos roam the savannah of Akagera National Park, and thriving patchworks of crops cling to steep slopes. The pulse of the nation, however, is truly found in the nation's people, whether in rural towns, cities or the rapidly developing capital, Kigali. Rwandans radiate an unfathomable strength and endurance of the human spirit, and the beauty of it shatters preconceptions, just as their stories of the genocidal past can break hearts. Travels here are incredibly rewarding, taking visitors to new highs as well as leading them on an introspective trip into the depths of the human condition.

Best time to visit
Mid-May to mid-March, to avoid the long rains

Top things to see
- The eyes of a mountain gorilla meeting yours on the forested slopes of Volcanoes National Park
- An Intore dance performance at the National Museum in Butare, which also hosts one of Africa's best ethnographical exhibits
- Papyrus gonolek and other rare bird species in the rich Nyabarongo wetlands
- The evolving – and charming – capital Kigali

Top things to do
- Take a safari in Akagera National Park, once again home to the Big Five
- Kick back on the beach at Kibuye, before travelling (or kayaking) along Lake Kivu
- Contemplate humanity's darkest side at the Kigali Memorial Centre
- Track chimps, L'Hoest's monkeys and colobus troops in Nyungwe Forest National Park

Getting under the skin
Read: *We Wish to Inform You That Tomorrow We Will be Killed With Our Families* by Philip Gourevitch, which delves into the horrors of the 1994 genocide
Listen: to *RWANDA, you should be loved* by the Good Ones; *Rwanda's Khashoggi: who killed the exiled spy chief?*, a podcast by *The Guardian*
Watch: *Gorillas in the Mist*, based on Dian Fossey's autobiography; *Hotel Rwanda*
Eat: grilled *tilapia* (Nile perch)
Drink: *icyayi* (sweet, milky tea)

In a word
Muraho (Hello, in Kinyarwanda)

Celebrate this
Godelième Mukasarasi, a rural development activist, social worker and genocide survivor, who founded the NGO Sevota to support widowed women and their children in the aftermath of the genocide.

Random fact
Since 2019, drones have been used to deliver blood and other medical supplies to rural regions.

S **CAPITAL** BASSETERRE // **POPULATION** 54,149 // **AREA** 261 SQ KM // **OFFICIAL LANGUAGE** ENGLISH

Saint Kitts & Nevis

Both St Kitts and Nevis embrace a mellow life revolving around 'limin' (hanging out, drinking and talking). The sugar cane plantations of yesteryear are long gone, and the economy is fast being overtaken by tourism. St Kitts is larger and given to loud, boisterous celebration, with bustling Basseterre, the huge Port Zante cruise terminal and the party strip and resorts of Frigate Bay. Nevis is the tranquil little brother that is almost impossibly alluring: circumnavigating the island on a two-hour drive is one of life's meandering pleasures.

Best time to visit
Year-round, although the hurricane season (June to October) has more storms

Top things to see
- Basseterre, the capital of St Kitts: equally thriving and shambolic
- St Kitts' Brimstone Hill Fortress, a rambling 18th-century fort and a Unesco site
- Historic Charlestown, the small main town on Nevis with a mellow vibe that makes you want to settle back on a park bench
- The sunset from the industrial-chic beach lounge bar Salt Plage
- The 360-degree view from the top of Mt Nevis
- Frigate Bay on St Kitts, with a string of beach shacks serving drinks until dawn

Top things to do
- Ride one of the ferries linking the two islands
- Sip a cocktail as you trundle around the coast on the Kitts Scenic Railway
- Stroll through the Botanical Garden of Nevis
- Windsurf on Nevis' Oualie Bay, a world-class site
- Dive in Sandy Point Bay far below the ramparts on Brimstone Hill

Getting under the skin
Read: *Only God Can Make a Tree* by Bertram Roach: part love story, part historical romance, with plenty of insight into island life
Listen: to Christmas music; there is a strong local tradition of setting old chestnuts to calypso and other Caribbean beats
Watch: *A Rose Between Thorns*: a 2017 debut by filmmaker Nigel 'TruCapo' Lewis, the first feature film production to be shot on location in St Kitts using an exclusively all-local cast and crew
Eat: saltfish and coconut dumplings, the national dish, for breakfast with a side of spicy plantain
Drink: CSR (Cane Spirit Rothschild), a potent potion made from sugar cane and most often mixed with Ting, a grapefruit soda

In a word
Meono (I don't know)

Celebrate this
The narrow-gauge railway on St Kitts is the only one in the Caribbean and features viewing platforms and waiters with cocktails.

Random fact
The federation of the two islands forms the smallest nation in the western hemisphere.

1. A herd of wild goats gathers at Frigate Bay

2. Beach bars line Frigate Bay's party strip

3. Brimstone Hill Fortress offers insights into the tumultuous past of the former Caribbean colonies

Best time to visit
Enjoy perfect weather with plenty of other visitors from December to May; at other times, you get rain, humidity and solitude

Top things to see
- Pigeon Island, a former hangout for pirates with evocative names like Wooden Let de Bois
- The St Lucia parrot, a rainbow-plumed bird that only lives on the island
- Soufrière, an authentic fishing town that's welcoming yet unaffected by tourists
- The buzzing markets of Castries, the capital, which are windows into island life
- Gros Islet, a genial mix of loafers, Rastas and beach bums

Top things to do
- Dive in Anse Chastanet, a marine park that's also ideal for snorkelling
- Climb the jagged Pitons, two volcanic formations towering over the island
- Dip into a volcanic mud bath at the Sulphur Springs in Soufrière
- Join a St Lucia street party
- Catch the breeze with kite-surfing on the south coast

Getting under the skin
Read: Derek Walcott's *Collected Poems, 1948-1984*, an anthology by the St Lucian Nobel Prize winner
Listen: to the local version of the banjo called the bwa poye
Watch: various films that use St Lucia for palm-tree scenes, such as *Dr Doolittle* (1967), *Superman II* (1980), *White Squall* (1996)
Eat: 'saltfish and green fig', salted cod cooked with unripe banana
Drink: Piton, a locally brewed lager

In a word
Bon jou (Good day, in Kwéyòl, which the French Creole islanders sometimes use)

Celebrate this
Two Nobel Prize winners have hailed from St Lucia: in 1992 Sir Derek Walcott won for literature and in 1979 Sir Arthur Lewis was awarded the prize in economics. This means St Lucia is the home of the most Nobel laureates per capita in the world.

Random fact
St Lucia is the only country in the world to be named after a woman, St Lucia of Syracuse.

1. Locals gather beachside in Soufrière

2. The colours below reflect the colours above in St Lucia

3. Marigot Bay is a popular sheltered anchorage for yachts

S CAPITAL CASTRIES // POPULATION 166,637 // AREA 616 SQ KM // OFFICIAL LANGUAGE ENGLISH

Saint Lucia

The colour wheel is simple on St Lucia: rich green for the tropical land, pure white for the ring of beaches and brilliant blue for the surrounding sea. And, if you look closely, you can fill out the rainbow. Yellows, oranges and reds emerge once you spot the flowers in the lush forest and take in the jaunty little villages with their brightly painted homes. (Yellow also gets help from the many banana plantations dotting the hilly countryside.) Take time for all the pleasures of surf and sand while tasting the cultural stew of loud reggae, piquant food and rum-fuelled escapades. Away from the coast, there are hikes amid the towering limestone Pitons that are alive with the echoes of waterfalls.

S CAPITAL KINGSTOWN // POPULATION 101,145 // AREA 389 SQ KM // OFFICIAL LANGUAGE ENGLISH

Saint Vincent & the Grenadines

Caribbean fantasies converge on this collection of 32 islands at the south end of the Leeward Islands. Party like a rock star in the fabled $150,000-a-week estates on Mustique, hang out with reggae-loving locals and itinerant fishers on St Vincent, or live your own pirate fantasy amid Tobago Cays. In fact SVG (as it's known) might be the ideal place to finally live out that dream of owning a yacht, as you can avoid the commitment by simply renting one and lazing your way about the islands, letting the winds and your moods guide you from one perfect spot to the next. But don't worry if that's not in the budget, cheap ferries make exploring these uncluttered, idyllic islands a breeze.

Best time to visit
Most people arrive December to May but the wet summer months can be nice and uncrowded

Top things to see
- Tobago Cays, a five-spot of tiny islands that could be a model of Caribbean perfection
- Kingstown, the buzzing Vincentian capital with its maze of porticoed stone alleyways
- St Vincent's Windward Hwy, a seemingly scripted mix of wave-tossed shores, placid coves and pastel-hued villages
- Fort Charlotte, an 1806 Kingstown edifice with commanding views of a dozen islands
- Port Elizabeth on the sweep of Bequia's yacht-dotted Admiralty Bay

Top things to do
- Trek through montane and cloud forest to the crater-topped La Soufrière volcano
- Go Bequia-bound to buy a model boat, hit the beach or explore the quirky Moonhole community
- Hunt for treasure amid the myriad variations of coral reefs and shipwrecks
- Claim your sandy patch on Mayreau's Saltwhistle Bay, a beach so fine it needs a sixth star
- Cruise to the Tobago Cays and Mustique in luxury under the sails of the *Friendship Rose*

Getting under the skin
Read: about village life in St Vincent in Cecil Browne's *The Moon is Following Me*
Listen: to reggae, steel bands, and local boy made good, Kevin Lyttle
Watch: the *Pirates of the Caribbean* movies, which use SVG as a principal location
Eat: *bul jol:* saltfish with tomatoes and onions, served up with roasted breadfruit
Drink: the locally distilled Sunset Rum

In a word
Check it? (Do you follow what I'm saying?)

Celebrate this
SVG is proud of its culinary traditions and doesn't import much food. Virtually everything that is served and eaten is grown in the fertile volcanic soil or caught in the abundant seas.

Random fact
SVG is home to the oldest botanical gardens in the western hemisphere, which was founded in 1765.

1. Boats moored on the beach at Admiralty Bay

2. Kingstown is the chief port and commercial centre of the country

3. An artisan creates model boats

S

Samoa

Slow down, way down, to Samoan time. Hardly anything disturbs the balmy peace here except for the occasional barking dog or passing pickup truck. All the attributes of an island paradise are here – cascades, jungles and endless blue lagoons – but without the usual tourist hoopla. While resort experiences are few, authentic cultural experiences abound. Music is everywhere: exuberant drumming resounds through the *fiafia* dance nights, choral music emanates from churches on Sundays while Samoan hip-hop is played day and night. Comprising two entities – the islands of independent Samoa and the US territory of American Samoa – the Samoan Islands share a history of being one of the strongest cultural forces in the Pacific.

Best time to visit
Between May and October, the dry season when many major Samoan festivals are held

Top things to see
- The Robert Louis Stevenson Museum in the author's former home, lovely Villa Vailima
- The eerie desolation of the Saleaula Lava Field
- An edge-of-the-world sunset at Cape Mulinu'u
- A game of *kirikiti*, Samoan-style cricket where dancing is as important as catching the ball
- Gorgeous geometric *pe'a* (traditional Samoan tattoos)

Top things to do
- Gaze at the sky while floating in To Sua Ocean Trench, a giant, iconic swimming hole
- Drift through the spectacular coral colonies of the Palolo Deep Marine Reserve
- Bathe in the jungle pool at Afu Aau Falls before standing atop pyramidal Pulemelei Mound
- Soak up the island vibe while strolling around the island of Manono
- Stride along Tutuila's mountainous spine to the top of Mt Alava

Getting under the skin
Read: Gavin Bell's *In Search of Tusitala*, which retraces Robert Louis Stevenson's South Sea voyages; or *The Beach at Falsea*, by Stevenson himself
Listen: to locally grown Samoan hip-hop – Mr Tee or New Zealand-based Samoan artists King Kapisi and Scribe
Watch: *O Le Tulafale* (The Orator), the slow-burning 2011 drama shot on 'Upolu
Eat: local favourites such as *oka* (raw fish in lime juice and coconut milk) and *palusami* (taro leaves cooked with coconut cream)
Drink: a crisp, ice-cold Vailima, one of the best beers in the Pacific

In a word
Fa'a Samoa (The Samoan way)

Celebrate this
Samoa has produced a dizzying list of famous athletes in sports from wrestling and rugby to soccer and gridiron, as well as two famous actors of Samoan descent: Dwayne 'The Rock' Johnson and *Star Wars* star Jay Laga'aia.

Random fact
The blue-green vermicelli-shaped palolo reef worm emerges at the same time every year to mate; these salty treats are said to be a potent aphrodisiac.

1. Net fishing in the harbor of Apia
2. The fairy grotto of the To Sua Ocean Trench swimming hole
3. A half-buried church at Sale'aula Lava Fields, where five villages where smothered by the 1905 eruption of Mt Matavanu
4. Samoans are rugby mad, and excel at all levels of the game

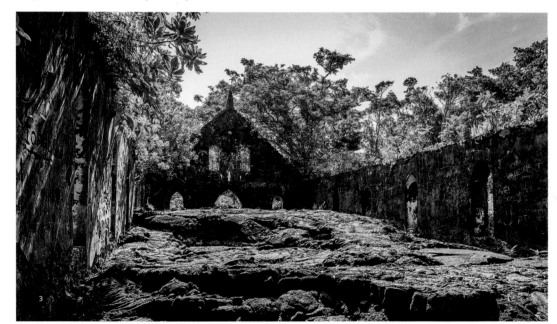

S

Best time to visit

May, June and September; July and August get crowded

Top things to see

- Palazzo Pubblico with its richly decorated facade
- The relics of St Marinus inside the neoclassical Basilica del Santo
- The changing of the guard in Piazza della Libertà, performed several times daily in summer
- Skinning devices, knee breakers and other ghastly torture devices at the Museo della Tortura

Top things to do

- Revel in pure unadulterated kitsch in San Marino's overdose of souvenir shops
- Take snaps of the republican soldiers, track down local euro coinage (or buy a souvenir set) and send a postcard home using a San Marino stamp
- Get your passport stamped at the tourist office – there are no border controls between San Marino and Italy
- Climb up two of the three towers pictured on the national flag, Guaita and Cesta (Montale is closed to the public), for sweeping coastal panoramas; on a clear day, you can even spy Croatia

Getting under the skin

Read: *A Freak of Freedom*, an evocative portrait of San Marino penned in 1879 by English explorer James Theodore Bent

Listen: to the sonatas and choral works of Cesare Franchini Tassini, the most famous of modern Sammarinese composers

Watch: Darryl Zanuck's *The Prince of Foxes* – the American director 'rented out' the entire republic to film the 16th-century period drama

Eat: *zuppa di ciliegie* – a 'soup' made from cherries, sugar and wine, served with local bread

Drink: a full-bodied Brugneto red, dry white Biancale or sweet dessert Oro dei Goti from San Marino's steeply terraced vineyards

In a word

Ciao (Hello/Bye)

Celebrate this

A stonemason turned monk turned saint, Marinus came from what is now Croatia to Italy, founding a monastery on Monte Titano in 301 CE and giving his name (in Italian) to the country he single-handedly founded.

Random fact

San Marino had to wait until 2004 to celebrate its first success in international football: a 1-0 defeat of Liechtenstein (this remains its only victory to date).

1. The oldest and largest of San Marino's castles, Torre Guaita dates to the 11th century

2. A cable car rises up Monte Titano from Borgo Maggiore to Città di San Marino

3. A guard stands on duty before San Marino's town hall

S **CAPITAL** CITTÀ DI SAN MARINO // **POPULATION** 34,232 // **AREA** 61 SQ KM // **OFFICIAL LANGUAGE** ITALIAN

San Marino

The world's oldest republic and Europe's third-smallest state, San Marino is a living quirk of history that today packs in the tourists. An island of stable sovereignty in the tempests of Italian politics since 301 CE, it has its own constitution (dating to 1600 – the world's oldest), army and currency, and attracts millions of visitors each year to the stunningly picturesque Città di San Marino, its capital and a Unesco-protected medieval hill town atop 750m-high Monte Titano. Beyond this lie its stunning Appenine territories and Sammarinese *castelli* (communes) such as Serravalle, Borgo Maggiore and Domagnano.

1. Cocoa beans in a freshly cut cocoa pod

2. Cão Grande, a 663m rock tower rising from the southern hinterland

3. Local life; the two islands are home to the second-smallest population of any African nation, after the Seychelles

S CAPITAL SÃO TOMÉ // POPULATION 211,122 // AREA 964 SQ KM // OFFICIAL LANGUAGE PORTUGUESE

São Tomé & Príncipe

Floating off Africa's west coast in the Gulf of Guinea, this nation is a tale of two islands. On little-developed Príncipe, a paradisical setting is on full display: rainforests blanket rolling hills and backdrop beaches, tropical birds circle volcanic rock formations, and aquatic life swarms its shores. While São Tomé boasts similar attributes, its crumbling colonial architecture, broken roads and decaying plantation homesteads illuminate the nation's current economic woes, as well as its troubled past. Uninhabited when discovered by Portuguese seafarers in 1470, the islands were populated by enslaved Africans and *degredados* (undesirables sent from Portugal), all of whom were forced to work on plantations. Their descendants, the São Toméans, gained a peaceful independence in 1975 and have thrived ever since.

Best time to visit
June to September (the dry season)

Top things to see
- The ocean lapping one of the world's most astounding beaches, Praia Banana
- Turtle hatchlings poking their noses out of the sand and hurrying down the beach to the sea
- Cão Grande, a massive phallic-looking volcanic tower rising from the jungle floor
- Pristine reefs, waters and prolific aquatic life

Top things to do
- Stand up paddle boarding with a volcanic skyline as a backdrop at Baía das Agulhas, Príncipe
- Savour each and every sip of your *bica* (tiny cup of coffee) at a São Tomé cafe
- Climb through forests filled with orchids and rare birds to summit Pico São Tomé
- Gain insight into the nation's planation history at Roça Agua Izé, where cacao is still grown
- Taste locally produced delights at the Claudio Corallo Chocolate Factory in the capital

Getting under the skin
Read: Miguel Sousa Tavares' novel *Equador* (Equator), a story about the governor of the islands' arriving from Portugal
Listen: to *Vôa Papagaio, Vôo!*, the seminal work of Gilberto Gil Umbelina
Watch: *Extra Bitter: The Legacy of the Chocolate Islands*, Derek Vertongen's exploration of slavery
Eat: *calulu* (smoked fish with a sauce of oca leaves, palm oil, chillies and herbs); São Toméan chocolate
Drink: coffee brewed from São Toméan beans

In a word
Lévé lévé (Easy, easy) – it's a mellow 'hello', and the motto of São Tomé

Celebrate this
The poetry of São Tomé's Conceição Lima.

Random fact
It was on the island of Príncipe where Albert Einstein's theory of relativity was proven in an experiment by Arthur Stanley Eddington on 29 May 1919.

Best time to visit
November to March

Top things to see
- Madain Saleh's ornate Nabataean tomb-facades chiselled into orange-hued rock outcrops
- Jeddah, for its old quarter brimming with preserved coral-block architecture and souq
- The isolated cliff-clinging villages and juniper-forest-clad ridges of Asir National Park
- Jubbah's cache of pre-Islamic rock carvings with petroglyphs dating back to 5500 BCE
- The Hejaz Railway's old stations and tracks made famous by TE Lawrence during the Arab Revolt

Top things to do
- Dive the Red Sea's Abu Madafi and Seven Sisters reefs with scarcely another diver in sight
- Venture into the dunescape of the Rub' Al-Khali (Empty Quarter)
- Boat trip to Farasan Island for beaches and exploring its pearl-merchant heritage
- Journey back to the 19th-century origins of Wahhabi Islam and the Al-Saud ruling dynasty in Diraiyah
- If you're Muslim, fulfil your religious duty and make the pilgrimage to Mecca

Getting under the skin
Read: *Adama* by Turki Al-Hamad; and *Cities of Salt* by Abdelrahman Munif
Listen: to Abdou Majeed Abdullah, Saudi Arabia's most enduring star of Arabic Pop
Watch: *The Perfect Candidate* directed by Haifaa Al-Mansour; or her debut, the critically acclaimed *Wadjda*
Eat: *khouzi* (lamb stuffed with a chicken that is stuffed with rice, nuts and sultanas)
Drink: cardamom-flavoured coffee

In a word
Allahu akbar (God is Great)

Celebrate this
Saudi Arabia launched its tourist e-visa scheme (for 49 countries) in 2019, opening up non-pilgrimage and non-business independent travel opportunities in the country for the first time.

Random fact
During Bedouin feasts, excessive conversation among the normally chatty Bedouin is considered a sign of poor manners.

1. Retractable canopies shade the courtyard outside the Prophet's Mosque in Medina

2. Camels, here at Riyadh's Janadriyah Festival, remain an important part of everyday Saudi life

3. Qasr Al-Bint at Madain Saleh is a cluster of rock-cut tombs and part of an ancient Nabataean settlement

S · CAPITAL RIYADH // POPULATION 34.1 MILLION // AREA 2,149,690 SQ KM // OFFICIAL LANGUAGE ARABIC

Saudi Arabia

Both birthplace of Islam and bastion of conservative values, Saudi Arabia has recently been all about change. After years of shunning non-pilgrimage tourism, Saudi is now encouraging all visitors, making its Nabataean ruins, desert dunes and Red Sea diving finally within travellers' grasp. Despite shrugging off its reclusive stance, the kingdom remains shrouded in controversial contradictions. Trumpeted societal reforms, an outward-looking new economic vision and Saudi's influential regional leadership role are shadowed by abuses of human rights.

S **CAPITAL** EDINBURGH // **POPULATION** 5.5 MILLION // **AREA** 78,772 SQ KM // **OFFICIAL LANGUAGES** ENGLISH & GAELIC

Scotland

A beguiling mix of sophisticated cities and brooding landscapes, no place quite eats into the traveller's soul like Scotland. True, the weather – buckets of rain and wind-whipped clouds – is hardly Mediterranean, but hike beneath castle-crowned crags and cinematic skies mirrored in lonely lochs, and it doesn't matter. This is one of Europe's last great wildernesses, its peaks the highest in the UK and its coastline fringed by islands in choppy waters where seals, dolphins and whales are a common sight. Cosmopolitan capital Edinburgh is culture-rich with wonderful museums, festivals and arts; traditional rival Glasgow stands its ground with innovative architecture, dining and nightlife. Round off your exploration with a nip of fine Scottish malt whisky.

Best time to visit
May to September; August for the Edinburgh festival season

Top things to see
- The fine view of the Firth of Forth from Edinburgh Castle
- Ben Nevis, high point of the Scottish Highlands
- Skara Brae, a Neolithic stone village built around 5000 years ago in Orkney
- The deserted beaches and walking trails of the remote Outer Hebrides
- Festival antics: Edinburgh's Hogmanay and Fringe and Shetland's Up Helly Aa are favourites

Top things to do
- Get a hole in one on the world's oldest golf course in St Andrews
- Hike spectacular cliff tops on the northern islands of Orkney
- Catch salmon, visit castles and enjoy lazy forest walks in Royal Deeside – the British Royal family has their country retreat, Balmoral, here
- Ride the Jacobite Steam Train through some of the country's finest scenery
- Get a taste of Glasgow's vibrant music scene at King Tut's or the Barrowland

Getting under the skin
Read: Alasdair Gray's *Lanark*, a dark, passionate fantasy about Scotland, Glasgow and storytelling; the work of national poet Robert Burns
Listen: to *Scotland the Brave* on the bagpipes; the Corries for authentic Scottish folk
Watch: Danny Boyle's *Trainspotting*, about Edinburgh's druggie underworld; *Whisky Galore*, a fact-based tale of a remote island's inhabitants finding a shipwreck full of the 'water of life'
Eat: haggis with neeps 'n' tatties (turnips and potatoes); cranachan (oats, cream, raspberries and whisky)
Drink: a dram of single-malt whisky; Irn-Bru, Scotland's radioactive-orange-coloured soft drink

In a word
Slàinte mhath (Cheers!)

Celebrate this
Edinburgh-born Sir Walter Scott achieved literary success in the nineteenth century and is credited with creating both the historical novel and the image of his homeland now famous worldwide (he reintroduced tartan 80 years after it had been banned).

Random fact
Nessie, the Loch Ness Monster, came to fame with photos in the 1930s, but its first mention is in the tale of 6th-century missionary St Columba.

1. On one of the great railway journeys of the world, the Jacobite steam train crosses the Glenfinnan viaduct

2. The monument to Scottish philosopher Dugald Stewart overlooks Edinburgh

3. Kilts fly in traditional Scottish Highland dancing

4. Red deer are a common sight amid the wild beauty of the Torridon region

S CAPITAL DAKAR // POPULATION 15.7 MILLION // AREA 196,722 SQ KM // OFFICIAL LANGUAGE FRENCH

Senegal

Although almost hewn in two by the borders of The Gambia, Senegal is a role model of political stability. Yet this devout West African country is far from dull, with a scintillating music scene that sets the beat for the nation's heart. Nowhere is this more evident than in Dakar, a spirit-filled capital that resonates with energy. Historical elements within Senegal also strike a chord: poignant monuments to slavery, the fusion of African and French culture, and Islamic sites of the Sufi Brotherhood. The landscape – ranging from coastal beaches and mangrove forests to savannah woodland and the dusty Sahel of the interior – is alluring too, particularly as it attracts millions of migratory birds en route between Africa and Europe.

Best time to visit
November to February – the dry season and relatively cool

Top things to see
- The tranquil Île de Gorée, with its compelling monuments to Africans cast into slavery
- Millions of migratory birds in the Parc National des Oiseaux du Djoudj
- Touba, home to the Mouride Sufi brotherhood and the extraordinary Grand Mosque
- Cap Skiring, among West Africa's most beautiful beaches
- Hippos, lions and dozens of other species in the Unesco-listed Parc National du Niokolo-Koba

Top things to do
- Tap into West Africa's music capital at the nightclubs of Dakar
- Glide through the shallows and mangroves of the Siné-Saloum Delta in a pirogue
- Sink into the sublime beats of the St-Louis Jazz Festival, one of the continent's best music events
- Explore the tropical landscapes and riverine world of the Casamance region

Getting under the skin
Read: *God's Bits of Wood* by Ousmane Sembène, a classic tale of colonial West Africa; Mariama Bâ's *So Long a Letter*, a window into the world of Senegalese women; or Anna Badkhen's *Fisherman's Blues*, a look at Senegal through its fishing community
Listen: to anything from Youssou N'Dour, one of the world's greatest singers
Watch: *Moolaade*, a tale about female circumcision; *Atlantics*, a modern story of refugees, ghosts and love, which was up for the Palme d'Or at the 2019 Cannes Film Festival
Eat: *tiéboudieune* (rice cooked in tomato sauce with chunks of fish, vegetables and spices)
Drink: hibiscus *bissap* (hibiscus juice); *bouyi* (baobab juice)

In a word
Asalaa-maalekum (Greetings, peace – in Wolof)

Celebrate this
The living legend of *mbalax*, singer Youssou N'Dour.

Random fact
At a time when other world leaders are changing constitutions to both increase their breadth of power and their time in office, Senegal's has recently done the opposite – new constitutional reforms have cut presidential term lengths to five years (from seven) and have limited their number to two.

1. Senegalese wrestling is the country's national sport
2. A market stall in Casamance
3. Domes of the Grand Mosque of Touba
4. Brightly coloured pirogues line the beach at Yoff

S CAPITAL BELGRADE // POPULATION 7 MILLION // AREA 77,474 SQ KM // OFFICIAL LANGUAGE SERBIAN

Serbia

Former Yugo-mates Croatia and Montenegro may have nabbed the coastlines, but for history, culture and feasting, landlocked Serbia is hard to beat. From the rollicking nightclubs of Belgrade to the hushed medieval monasteries in the south, the green slopes of Tara mountain to the Danube's Iron Gates gorge, endearing village cottages to mountaintop communist memorials, this is a land which embraces – and revels in – its striking contrasts. The people have toughed out many political and economic challenges but they know how to enjoy life, with farm-to-table cuisine, packed cafes, unhurried street life and boisterous music festivals. EU bound and on the up, Serbia is a destination to watch.

Best time to visit
Between May and September

Top things to see
- Kalemegdan Citadel in Belgrade, over the confluence of the Danube and Sava
- The twists and turns of the Šargan Eight narrow-gauge railway on Tara mountain
- Art nouveau architectural treasures in Hungarian-influenced Subotica
- The ancient frescoes of Studenica Monastery, one of the most sacred sites in Serbia
- Griffon vultures soaring above the incredibly green, zigzagging Uvac Canyon

Top things to do
- Party until dawn (and beyond) at the EXIT festival in Petrovaradin Fortress in Novi Sad
- Taste the native Prokupac variety in the 19th-century wine cellars of Rajac and Rogljevo villages
- Hit the piste at Kopaonik, Serbia's premier ski resort
- Release your inner nerd with sci-fi-ish interactive elements at the Nikola Tesla Museum in Belgrade
- Cycle the Eurovelo 6 along the Danube or join a boat cruise through the formidable Iron Gates gorge

Getting under the skin
Read: *Destiny, Annotated* by Radoslav Petković, the most awarded novel in the history of Serbian literature
Listen: to *trubači*, wild, haunting brass sounds influenced by Turkish melodies and Austrian military music
Watch: the films *Premeditated Murder* and *Thunderbirds*, for a great snapshot of the 1990s in the Serbian capital
Eat: *ćevapčići* (grilled kebab) or *pljeskavica* (spiced beef patties)
Drink: *rakija*, a fiery spirit made from fermented fruit – particularly *šljivovica*, made of plums

In a word
Živeli (Cheers!)

Celebrate this
This small nation has several sports heroes who made it big on the international scene despite the odds, from former Manchester United captain Nemanja Vidić to NBA star Nikola Jokić, but no one can match the achievements of tennis world number-one Novak Djoković, with 17 Grand Slam titles to his name to date.

Random fact
The Serbs use both the Cyrillic and Latin alphabets, switching between them without a second thought.

1. Belgrade's fortress in Kalemegdan Park has views of the city and river

2. A snowy scene on Tara mountain

3. A Balkan brass band prepares for the annual Guča Trumpet Festival

Best time to visit
March to May and September to November

Top things to see
- Anse Lazio's perfect sands, palms and waters, punctuated by sculpted granite monoliths
- The rare and ever-so-suggestive coco de mer palms growing in primordial Vallée de Mai on Praslin Island
- Thunderous clouds of feathers flocking in the skies over Bird Island
- Island-topping vistas following a trek in the jungle-clad hills of Morne Seychellois National Park
- A perfectly prepared plate of succulent seafood being placed in front of you

Top things to do
- Wade, boulder and bushwhack to reach Anse Marron on La Digue, the world's most beautiful natural pool and beach combo
- Paddleboard atop the azure waters off Anse Volbert
- Forget *Jaws* and share the depths diving with sea creatures big and small at the not-so-aptly named Shark Bank
- Satisfy any *Robinson Crusoe* fantasies on Alphonse Island
- Swim, picnic and greet giant tortoises on Curieuse Island

Getting under the skin
Read: *Seychelles Since 1770: History of a Slave and Post-Slavery Society* by Deryck Scarr
Listen: to Jean-Marc Volcy, who's fused modern Creole pop with traditional folk music
Watch: *Le Monde de Silence*, Jacques Cousteau's ground-breaking documentary, much of which was filmed at Seychelles' Assomption Island
Eat: *trouloulou* and *teck teck*, two local varieties of shellfish
Drink: *calou* (a palm wine that will put a bounce in your step)

In a word
Bonzour (Good morning, in Kreol Seselwa)

Celebrate this
Rodney Govinden, Seychelles' top sailor; having already represented his nation at the 2016 Summer Olympics in Rio de Janeiro, he subsequently qualified in the Men's Laser event for the 2020 Summer Olympics in Tokyo.

Random fact
Weighing up to 20kg, the seed in the famously erotic fruit of the Seychelles' female coco de mer palm is the plant kingdom's largest and heaviest.

1. Fishing the emerald waters off La Digue

2. Aldabra Atoll is home to about 150,000 giant tortoises

3. Anse Source d'Argent, straight out of any beach-lover's fantasies

CAPITAL VICTORIA // **POPULATION** 95,981 // **AREA** 455 SQ KM // **OFFICIAL LANGUAGES** CREOLE, ENGLISH & FRENCH

Seychelles

Almost 700 million years in the making, the islands of the Seychelles have been painstakingly shaped into utter tropical perfection by Mother Nature. Whether hewn from granite or grown from corals, these masterpieces are all fringed with white-sand beaches, alluring waters and palms laden with exotic fruit. Looking to the future, the government has recently taken steps to protect and preserve almost 210,000 sq km of the surrounding Indian Ocean, almost a third of the nation's exclusive economic zone. All 115 of these Indian Ocean islands were uninhabited until the 18th century – many still are. The burgeoning society is primarily African in origin, though it's infused with French, British, Indian, Chinese and Arab influence.

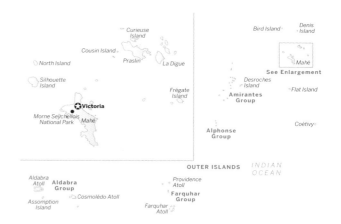

Best time to visit
November to April

Top things to see
- Haunting Bunce Island, an important landmark in the tragic history of slavery
- Turquoise waters meeting golden sands on the Turtle Islands, a slice of rarely visited paradise
- The summit of Mt Bintumani, West Africa's second-highest mountain, after a well-prepared and executed climb
- Intricately patterned butterflies flitting through shafts of light deep within Gola Rainforest National Park
- The bubbling energy, colour and charm emanating from Freetown, the nation's capital

Top things to do
- Learn about local communities and culture while walking the Tiwai Heritage Trail
- Canoe through the chatters and songs of monkeys and birds in the surrounding forest in Outamba-Kilimi National Park
- Tuck into barbecued lobster that has just been pulled from the sea on one of Freetown's beaches
- Enter a slab of ancient rainforest at night to see the elusive pygmy hippo on Tiwai Island
- Understand the plight (and marvellous nature) of our closest relatives at Tacugama Chimpanzee Sanctuary

Getting under the skin
Read: *A Long Way Gone: Memoirs of a Boy Soldier* and *Radiance of Tomorrow*, both by Ishmael Beah
Listen: to palm wine music (or *maringa* as it's known locally), whose finest exponent was the late SE Rogie
Watch: the disturbing documentary *Cry Freetown* by Sorius Samura; or *Ezra*, the award-winning movie about an ex-child soldier trying to pull the pieces of his life together
Eat: *punky* (a thick stew of onion, chilli peppers, squash, oil, spices and fish or meat) served on rice
Drink: *bissap*, brewed hibiscus flowers with vanilla and sugar

In a word
Owdibody (How are you? – literally 'How's the body?')

Celebrate this
Michaela DePrince, a professional ballet dancer, goodwill ambassador for War Child and author of *Taking Flight: From War Orphan to Star Ballerina*.

Random fact
On Valentine's Day in 1972 miners in the Koidu area discovered the Star of Sierra Leone, a 968.9-carat diamond – it still ranks as the largest-ever alluvial diamond discovered.

1. Keeping shop in a village near the capital, Freetown

2. Strung between the mountains and the sea, Freetown is a cheeky, quicksilver city

3. Gola Rainforest National Park is a locus of conservation, ecotourism and national pride

S CAPITAL FREETOWN // POPULATION 6.6 MILLION // AREA 71,740 SQ KM // OFFICIAL LANGUAGE ENGLISH

Sierra Leone

Sierra Leone deserves more than its past. So now, with peace and renewed health, the future of its hopes may just be there for the taking. Nestled into the coastline of West Africa, Sierra Leone's landscape is as rich as the diamonds and other mineral resources it holds deep within: mountains fill the horizon, tropical islands punctuate the waters, soft sands stretch uninterrupted along the coast and blankets of rainforest hold rare wildlife. These natural spectacles have the power to captivate visitors, as do interactions with the incredible people who call this nation home. And in that regard, responsible tourism can have a rewarding role to play as Sierra Leoneans write their next chapter.

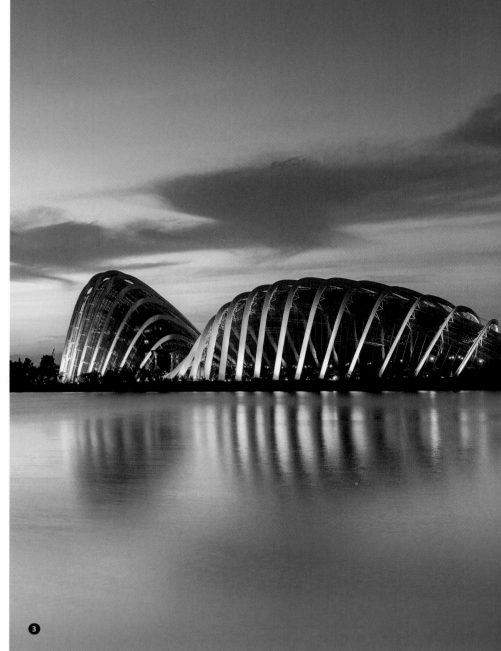

Best time to visit

Year-round – February to October tends to be drier, but skies are hazy from June to October

Top things to see

- Gardens by the Bay, Singapore's most show-stopping green space
- History and culture at the National Gallery Singapore and Asian Civilisations Museum
- Temples, mosques and cosmopolitan cafes and bars among the laneways of Chinatown, Little India and Kampong Glam
- Tiong Bahru, a revived 1930s housing estate, packed with cafes and boutiques
- A menagerie of exotic beasties at the Singapore Zoo and Night Safari

Top things to do

- Light a joss stick at the historic Thian Hock Keng, Singapore's oldest Hokkien temple
- Take a bumboat to Pulau Ubin for some monkey and monitor-lizard spotting in a natural setting
- Feast your way around the Straits in one of Singapore's legendary hawker centres
- Negotiate the jungle canopy along the TreeTop Walk at MacRitchie Reservoir
- Join in the fun of one of Chinatown's colourful festivals

Getting under the skin

Read: *Foreign Bodies*, Hwee Hwee Tan's gripping tale of young people on the wrong side of the Singapore justice system
Listen: to local and international DJs man the decks at iconic Singapore nightclub Zouk
Watch: Woo Yen Yen and Colin Goh's *Singapore Dreaming* or Tay Teck Lock's *Money No Enough* for insights into the Singapore psyche
Eat: Peranakan Straits Chinese cuisine at Singapore restaurants including Candlenut and Blue Ginger
Drink: Tiger beer (the national brew), *kopi* (strong coffee with condensed milk), *teh tarik* (strong, sweet 'pulled' tea with condensed milk)

In a word

Kiasu (Fear of losing) – a Hokkein/Singlish term for one of the defining traits of the competitive inhabitants of Singapore

Celebrate this

With Chinese, Malay and Indian cultures all contributing to Asia's most diverse country, centuries of trading and historical interaction have blessed Singapore with three shared and equally celebrated strands of national identity.

Random fact

Singapore is the world's largest exporter of exotic aquarium fish.

1. Colourful shophouse facades in historic Chinatown

2. Chinatown's restaurant and bar scene is booming

3. High-tech Mediterranean and cloud-forest biodomes at Gardens by the Bay

4. At night, Gardens by the Bay's Supertrees burst into light for the Garden Rhapsody show

Singapore

Some travellers knock Singapore for its corporate mindset, draconian laws and high prices, but this Southeast Asian metropolis can no longer be accused of lacking redeeming features. Fans of the city-state rave about its green spaces, great shopping, fabulous food and an intoxicating blend of Indian, Chinese and Malay culture. With a vibrancy that now competes with that of Bangkok and Hong Kong, there's plenty to see, from quirky local neighbourhoods and world-class museums and galleries to historic temples where the air is thick with incense. The city is fabulously well organised, and getting around to experience Singapore's cosmopolitan energy is easily achieved on super-efficient public transport. More than a stopover before further Asian travels, Singapore is also a compelling destination in itself.

S CAPITAL BRATISLAVA // POPULATION 5.4 MILLION // AREA 49,035 SQ KM // OFFICIAL LANGUAGE SLOVAK

Slovakia

This small and largely mountainous nation at the heart of Central Europe is still recognised by much of the outside world as the less famous half of the act formally known as Czechoslovakia. Since independence, Slovakia's capital, Bratislava, has been drawing visitors with a hilltop castle and charming old town, but nature is still king. The great outdoors include the High Tatras mountains, which attract passionate hikers and skiers, and the meadows, forests, rocky gorges and waterfalls of Malá Fatra and Slovenský Raj National Parks. Unpretentious Gothic cities, stately castles, wooden churches and wine-producing villages complete the scene.

Best time to visit
May to September or December to March

Top things to see
- Bratislava's Old Town, with its resplendent castle overlooking the Danube
- Spiš Castle, Europe's biggest fortress
- Neat pastel facades on the Gothic-Renaissance burghers' houses in Bardejov
- The precipitous peaks and pine-topped ridges in the Malá Fatra National Park
- Prickly spires and battlements on Bojnice Castle, the most visited chateau in Slovakia

Top things to do
- Plunge into a thermal pool, breathe 'seaside' breezes in a salt cave, or be wrapped in hot mud at Piešťany spa resort
- Dip your toes over the edge of a *plte* (wooden raft) down Dunajec Gorge
- Clamber up the ladder and chain ascents to the precipice in Slovenský Raj National Park
- Crunch through the snow on the walking trails of the High Tatras

Getting under the skin
Read: the straightforward tales of feisty Slovakian women in *That Alluring Land: Slovak Tales* by Bozena Slancikova-Timrava
Listen: to wailing *gajdy* (bagpipes) and *fujara* (shepherd's flutes) that are central to much Slovakian folk music
Watch: internationally acclaimed *Krajinka*, directed by Martin Sulik: 10 vignettes of Slovakian rural life, landscape and ways throughout the 20th century
Eat: *bryndzové halušky* (potato dumplings with sheep's cheese and bacon)
Drink: local beers such as dark, sweet Martiner or full-bodied Zlatý Bažant; or quaffable local wines

In a word
Ahoj (Hello)

Celebrate this
Folk traditions still hold sway in Slovakia. The largest *skanzen* (open-air museum), the Museum of the Slovak Village in Martin, also features the country's only museum dedicated to Roma people.

Random fact
Venus of Moravany, a headless female fertility symbol carved from mammoth bone found near Piešťany in 1938, is almost 25,000 years old.

1. Standing on the site of an 11th-century castle, Bojnice Castle was designed in the 19th century to resemble a French chateau

2. Bratislava's main square hosts festive markets

3. Eurasian lynx live in Malá Fatra National Park but they're elusive; look for footprints in snow

4 Autumnal foliage arrives in Slovenský Raj National Park

Best time to visit
May to September

Top things to see
- The view over Ljubljana's Old Town and the bridges of the Ljubljanica River from the ramparts of Castle Hill
- Subterranean chambers and cave-dwelling salamanders in Postojna
- Shimmying dancers in shaggy sheepskin and masks at the Kurentovanje festival
- The sparklingly azure Adriatic Sea at Piran, with its Venetian ambience
- The snow-white horses of Lipica, bred for the Spanish Riding School in Vienna

Top things to do
- Ring the wishing bell in the Church of the Assumption on Bled Island, then return to shore in a piloted gondola
- Hike between mountain huts on well-marked trails in the Julian Alps
- Raft the foaming waters of the Soča River
- Sip wine at the source in Maribor, home to the world's oldest living grapevine
- Shop for remarkably detailed lace in former mercury-mining town Idrija

Getting under the skin
Read: *Forbidden Bread* by Erica Johnson Debeljak, a memoir of an American woman coming to terms with life in Slovenia
Listen: to traditional folk 'big band' music, featuring pan pipes and zithers; or the electro-industrial stylings of Laibach, infamous for being the first Western group to perform in North Korea
Watch: Damjan Kozole's *Rezervni Deli* (Spare Parts), a provocative and award-winning tale of the trafficking of illegal immigrants through Slovenia
Eat: *žlikrofi* (dumplings filled with cheese, bacon and chives), followed by *štruklji* (sweet, cottage-cheese dumplings) or *palačinke* (pancakes)
Drink: wine such as peppery red Teran; or *žganje* (fruit brandy distilled from many fruits)

In a word
Zdravo (Hello)

Celebrate this
Mt Triglav is a source of Slovenian pride: along with connections to folk tales and WWII resistance fighters, it's also the country's highest peak and climbing its three summits is a national rite of passage.

Random fact
A national icon of Slovenia is the *kozolec* (hayrack) – there's now a museum in Šentrupert entirely dedicated to the craftsmanship of these well-loved farming tools.

1. Looking over Piran and the Adriatic from the city's walls; its main square is named after composer Giuseppe Tartini who was born here

2. Slovenia's Soča River offers some of the best white-water action in Europe

3. See Ljubljana from a paddleboard as the Ljubljanica River flows sedately through the Slovene capital

4. Pericnik waterfall is one highlight of Triglav National Park in the northwest; Slovenia is noted for its natural spaces and activities

Slovenia

S   **CAPITAL** LJUBLJANA // **POPULATION** 2.1 MILLION // **AREA** 20,273 SQ KM // **OFFICIAL LANGUAGE** SLOVENIAN

Contrasts abound in tiny Slovenia. A modern, forward-looking nation where myths of a three-headed mountain god live on and nostalgia for the agrarian past remains strong. A thoroughly Slavic country yet with obvious Italian and Austro-Hungarian influences. A place with Alpine peaks where snow may last into summer, but where Mediterranean breezes may suddenly raise temperatures. With almost half of its total area covered in forest, this is one of the greenest countries on earth, something in which its residents take immense pride. Villages are orderly, churches picturesque, the capital pocket-size but perfect and castles imposing, yet the pagan spirit of the people lives on in their raucous and colourful festivals.

Best time to visit
June to September has mild weather – good for hiking but rough seas mean diving conditions aren't ideal

Top things to see
- Rusty WWII relics around Honiara
- The artificial stone and coral islands of Malaita's Langa Langa Lagoon
- The sensory overload of Honiara's central market
- Leaf-mound nests of megapodes, birds that use volcanic heat to incubate their eggs
- Eerie skull caves, the final resting places of vanquished warriors and chiefs

Top things to do
- Dive the fantastic 'Iron Bottom Sound' WWII wrecks off Guadalcanal
- Assist rangers tagging sea turtles on ecofriendly Tetepare Island
- Take a dip in the natural pools beneath Mataniko or Tenaru Falls
- Kayak, dive or snorkel through the marine-life-rich waters of Marovo Lagoon
- Surf the crowd-free point breaks off Pailongge on Ghizo

Getting under the skin
Read: *Solomon Time*, Englishman Will Randall's funny, insightful account of starting a chicken-farming business on Rendova; or Graeme Kent's Kella & Conchita Mystery series, set in the Solomons
Listen: to Onetox, a groundbreaking Solomon-reggae band, or Narasirato, a panpipe ensemble mixing traditional Malaita sounds with contemporary beats
Watch: Terrence Malick's *The Thin Red Line*, about the WWII battle for Guadalcanal; or 2018's documentary *Liborio*, a visual feast set in Malaita
Eat: *ulu* (breadfruit), the Solomon Islands' staple
Drink: SolBrew pale lager, the local brew

In a word
No wariwari (No worries)

Celebrate this
Beautiful, biodiverse Marovo Lagoon is the world's biggest saltwater lagoon. It's been nominated for World Heritage status; if granted, it will join East Rennell – the planet's largest raised coral atoll – on the list, currently the Solomons' only entry.

Random fact
There are 67 indigenous languages in the Solomons; though English is the official language, Pijin is used in day-to-day communication.

1. Eclectus parrots are native to the Solomons and neighbouring islands; males are green, females are red

2. A tropical climate of high temperatures and regular rainfall means foliage is lush

3. The MS *World Discoverer* cruise ship roams no more, lying aground in Roderick Bay since 2000

S   **CAPITAL** HONIARA // **POPULATION** 685,097 // **AREA** 28,896 SQ KM // **OFFICIAL LANGUAGE** ENGLISH

Solomon Islands

Want to get off the beaten path? That's easy: there is no beaten path in the Solomon Islands. It's just you, the ocean, dense rainforests and traditional villages, and it feels like the world's end. With a history of headhunting, cannibalism and (more recently) civil unrest, the islands are much safer than their past suggests and are now an ecoadventurer's dream. The volcanic, jungle-cloaked islands jut up dramatically from the Pacific and are surrounded by croc-infested mangroves, huge lagoons, beaches and lonely islets. Solomon Islanders are laid-back, friendly and still practise ancient arts and till their village gardens the way they have for thousands of years.

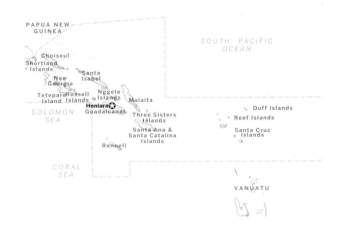

CAPITAL MOGADISHU // **POPULATION** 11.8 MILLION // **AREA** 637,657 SQ KM // **OFFICIAL LANGUAGE** SOMALI

Somalia

A problematic patchwork, this nation perched on the Horn of Africa is made up of three seemingly insoluble elements: Somalia, Puntland and Somaliland. The former has dominated the press for more than 30 years, with armed conflict, warlords and Mogadishu hotel bombings, while Puntland (autonomous since 1998) became synonymous with the piracy that plagued the Indian Ocean between 2007 and 2013. The golden child had been Somaliland, a self-proclaimed republic since 1991. With a record of maintaining both peace and order, it lured bold travellers with astounding archaeological sites, epic beaches and the welcoming Isaq clan of Somalis. But Somaliland's run in the sun ended when the risk of terrorism and kidnap carried over from the other two in the past decade, making it also unsafe.

Best time
December to March, when it is coolest

Places of interest
- The treasures of Las Geel, caves containing hundreds of well-preserved Neolithic rock paintings
- Ras Xaafun, which provides an incredible vista from the Horn of Africa
- Merca, a fascinating Arab coastal town with myriad alleyways
- Sheekh, the site of a 13th-century necropolis
- The white sands of Baathela Beach near Berbera
- A wealth of aquatic life off the Bajuni Islands

Local customs
- Bonfires and dancing at the annual festival of Neeroosh, held each July in celebration of the start of the solar year
- Widespread use of the narcotic leaf *qaat*, with some of those interested in Sufi spiritualism using it to reach a trance-like state as a way of communing with Allah
- Celebrating a vibrant musical heritage that focuses on Somali folklore

Getting under the skin
Read: Nuruddin Farah's 'Past Imperfect Trilogy': *Links* (1978), *Knots* (2006) and *Crossbones* (2011); *The World's Most Dangerous Place* by James Fergusson
Listen: to *The Journey* by Maryam Mursal, the first woman to sing Somali jazz (Peter Gabriel sings back-up in this effort)
Watch: *Men in the Arena*, a story of two Somali football stars chasing their dreams
Eat: *anjeero* (local flatbread) topped with sheep liver and onions
Drink: *shaah* (Somali black tea) with *heel* (cardamom) and *qarfe* (cinnamon)

In a word
Ma nabad baa? (A greeting, meaning 'Is it peace?')

Celebrate this
Hamza Sheikh Hussein, a Somali MP who has been an activist for women in national leadership: she also campaigns against female genital mutilation.

Random fact
At its height piracy generated nearly US$200 million in annual income, making it Somalia's biggest industry.

1. Looking out over Somaliland from the Las Geel caves

2. Paintings in the Las Geel caves are mostly of cows in vivid colours and many seemingly with decorative neckwear; dogs and humans are also represented

3. A queue for food on the first day of Eid al-Adha in Mogadishu

1. The reason for the bright hues of Cape Town's Bo-Kaap quarter is that housing was originally built for indentured workers who were unable to own their homes or change them; when that law changed, many got painting

2. Tugela Gorge makes for a great day hike in the Royal Natal National Park

3. White rhinos are protected in Sabi Sand game reserve, adjoining Kruger National Park

Ⓢ CAPITAL PRETORIA (ADMINISTRATIVE), BLOEMFONTEIN (JUDICIAL), CAPE TOWN (LEGISLATIVE) // POPULATION 56.5 MILLION // AREA 1,219,090 SQ KM // OFFICIAL LANGUAGES ZULU, XHOSA, AFRIKAANS, SEPEDI, ENGLISH, TSWANA, SESOTHO, TSONGA, SWATI, VENDA & NDEBELE

South Africa

The 'Rainbow Nation' moniker doesn't cover the half of it. The astonishing diversity of South Africa is not just seen in its people, but in everything. Landscapes and wildlife collide to offer incredible sights and activities – beachgoers share sand with penguins on the Cape's beaches, divers encounter whales (and great whites), safari cognoscenti watch iconic species in the bushveld, and hikers eye vultures soaring over the snowy peaks of the Drakensberg. The historical sites, ranging from the Cradle of Humankind to the Apartheid Museum, are no less compelling. And South Africa's human drama – its pain, its injustice and its hopeful spirit – is palpable. The result is sobering and challenging, fascinating and inspiring; alluring enough to keep most visitors returning time and time again.

Best time to visit
Year-round, with spring (September to November) and autumn (April to May) ideal almost everywhere

Top things to see
- The art and surrounds of the new Zeitz Museum of Contemporary Art Africa in Cape Town
- San cave paintings, bearded vultures and mountain vistas in the Drakensberg
- The Big Five and other iconic wildlife in Hluhluwe-iMfolozi Park
- Colourful carpets of spring flowers in Namakwa
- The galleries, microbreweries, cafes and surrounding mountain peaks of Clarens

Top things to do
- Self-drive safari in Kruger National Park
- Sample world-class wines, cycle through vineyards and dine in style, all within the Winelands
- Walk the cliffs and beaches of the Wild Coast
- Hike by day and stargaze by night in the wilderness of the Cederberg
- Embrace modern Jo'burg in Braamfontein, Newtown, 44 Stanley and the Maboneng precinct

Getting under the skin
Read: *Long Walk to Freedom* – Nelson Mandela's inspirational autobiography; *A History of South Africa* by Frank Welsh; *The Cape Town Book*, an impressive history by Nechama Brodie
Listen: to 'Nkosi Sikelel' iAfrika' (God Bless Africa) – part of the South African national anthem of unity
Watch: *Invictus*, which covers the historic events around the 1995 Rugby World Cup in South Africa
Eat: *biltong* (dried and cured meat); *mealies* (maize); and *boerwors* (sausages)
Drink: wines from the Cape Winelands

In a word
Howzit?

Celebrate this
A new generation of filmmakers, such as Bongiwe Selane, is telling fresh South African stories.

Random fact
2000 native species of plant live on Table Mountain.

Best time to visit
September to November, for spectacular autumn colours

Top things to see
- Markets, museums and medieval city gates in the fast-paced capital, Seoul
- Acres of tombs, temples and dynastic ruins in historic Gyeongju
- The royal mausoleums of the Baekje dynasty at Gongju and Buyeo
- The old-fashioned Korean way of life on the islands of Dadohae Haesang National Park
- Mountains, forests, hot springs, temples and serenity at Seoraksan National Park

Top things to do
- Stay overnight in a Korean temple, sharing vegan food and devotions with the resident monks
- Party with hipsters in Seoul's trendy Hongdae and Seongsu-dong neighbourhoods
- Ski and snowboard on Olympic-standard slopes at Alpensia and Yongpyong ski resorts
- Stand as close as you safely can to the world's most volatile border on a tour of Panmunjom and the Demilitarized Zone (DMZ)
- Hike around the stunning coastline of Jeju Island, following the scenic Jeju Olle Trail

Getting under the skin
Read: *A Geek in Korea* by Daniel Tudor; *I'll Be Right There* by Shin Kyung-sook
Listen: to the lyrical beauty of *pansori* – musical storytelling, often described as the Korean equivalent of the blues
Watch: Bong Joon-ho's award-winning social satire *Parasite*; Park Chan-wook's extreme *Old Boy*, or the same director's elegant *The Handmaiden*
Eat: *kimchi* (fiery pickled cabbage with chilli) and *galbi* (barbecued ribs) grilled at your table
Drink: *soju* (local rice spirit); or *bori cha* (warming tea made from roasted barley)

In a word
Jeong (Emotional attachment bordering on love)

Celebrate this
It's not just politicians trying to break the impasse between North and South Korea; civilians in the south regularly load balloons with everything from peace leaflets and money to rice and coronavirus masks and float them over the border.

Random fact
South Korea has an entire holiday dedicated to its writing system: Hangeul Day, celebrated on 9 October.

1. Traditionally attired musicians in South Gyeongsang Province playing the Korean *janggu* drum

2. Fresh fish for sale in the southern port city of Busan

3. The Gyeongbokgung Palace complex contains museums, gardens and amazing architectural sights

CAPITAL SEOUL // **POPULATION** 49 MILLION // **AREA** 99,720 SQ KM // **OFFICIAL LANGUAGE** KOREAN

South Korea

South Korea is a beacon of progress and modernity in Asia, second only to Japan in its enthusiasm for scientific advances and future tech. Korean expats have spread the cuisine of their homeland across the globe, but surprisingly few travellers explore this ultramodern but deeply traditional corner of Asia. While the public image of South Korea is all skyscrapers and smartphones, for every high-tech metropolis there's a medieval fortress or a tranquil national park where locals come to escape the 21st-century hubbub. South Korea's uneasy relationship with its neighbour to the North casts a periodic shadow but, for travellers, this is a place to experience Confucian culture without political baggage, enjoying every mealtime as a feast.

S · CAPITAL JUBA // POPULATION 10.6 MILLION // AREA 644,329 SQ KM // OFFICIAL LANGUAGE ENGLISH

South Sudan

Born out of more heartache and bloodshed than any other nation on the continent, South Sudan quietly came to be in 2011. The half-century of civil war – Africa's longest-running conflict – that preceded the independence referendum was fuelled by oil riches and pitted Islam against Christianity, Arab against Black African and central government against regional autonomy. Sadly, just 17 months after inception, civil war broke out, the door to travellers slammed shut and South Sudan remains one of the world's poorest and least developed nations. When access returns, visitors will encounter unique cultures, such as that of the pastoralist Dinka people, and natural attractions, including what may be the planet's largest animal migration.

Best time to visit
October to March (the dry season)

Places of interest
- The southern swamps, where 800,000 white-eared kob and half a million topi and Mongalla gazelles take part in one of the planet's largest mass movements of animals
- Bandingalo National Park, home to hippos and wild dogs
- The resting place of John Garang de Mabior (or Dr John, as he's known locally), a South Sudan rebel and former first vice-president of Sudan, who was instrumental in ending that country's civil war
- Nimule National Park, for its elephants, buffaloes and wild scenery
- Juba, a burgeoning capital city

Local customs
- There are many important traditional dances, such as the Dinka's leaping, the Beja's sword dance and the Halfa's *barabrah*
- Eating sweet Maridi honey, which is known for its healing properties
- Dinka people use cattle urine to wash, as well as to dye hair

Getting under the skin
Read: *First Raise a Flag: How South Sudan Won the Longest War but Lost the Peace* by Peter Martell
Listen: to Emmanuel Jal's *Ceasefire*, an album that calls for peace across his land, which he fittingly created with Abdel Gadir Salim, a singer and composer from northern Sudan
Watch: *Lost Boys of Sudan*, a documentary film by Megan Mylan and Jon Shenk, following two Dinka boys who escaped Sudan's civil war (and lion attacks) to live an altogether different life in America
Eat: *asida* (porridge made from sorghum) – it's served with a meaty sauce or vegetables
Drink: *aradeab* (tamarind juice)

In a word
Salaam aleikum (Peace upon you)

Celebrate this
Guor Marial, Santino Kenyi and Margret Hassan: the first three athletes to compete for South Sudan at an Olympics (2016 Summer Games in Rio de Janeiro).

Random fact
Bulls are so important in South Sudan that many people are named after them; white bulls (the most prized) are sacrificed for celebrations and peace.

1. Mundari herders, who live along the banks of the White Nile River, tend to their Ankole-Watusi cattle with exteme care; the animals are rarely killed for meat but provide many resources instead

2. A wheelchair basketball team of amputees practises for the Paralympics; some 400,000 South Sudanese have a disability due to war

3. Huts with compounds for livestock in Unity State, South Sudan

Best time to visit
Year-round in the south, May to September in the north

Top things to see
- The Alhambra, the exquisite highpoint of Andalucía's Islamic architecture
- Córdoba's Mezquita, perfection rendered in stone
- Gaudí's Barcelona, one man's astonishing legacy that came to define a city
- Madrid's golden triangle of art, three of the world's best art galleries
- Santiago de Compostela's cathedral, Spain's most sacred corner and a flight of architectural extravagance

Top things to do
- Go on a food crawl sampling San Sebastián's world famous *pintxos* (Basque tapas)
- Hike the Pyrenees in Catalonia or Aragón
- Drive along the dramatic Galician coastlines of Rías Altas or the Costa da Morte
- Laze on a secluded and perfect beach on Ibiza or Formentera
- Escape the modern world in whitewashed *pueblos blancos*

Getting under the skin
Read: the classic *Don Quijote de la Mancha* by Miguel de Cervantes; *The Shadow of the Wind* by Carlos Ruiz Zafón
Listen: to Camarón de la Isla, Paco de Lucia and Carmen Linares for flamenco; Manuel de Falla for classical music
Watch: any film by Pedro Almódovar, especially *Volver* (Return) or Oscar-winning *Todo Sobre Mi Madre* (All About My Mother)
Eat: tapas; wafer-thin slices of *jamón ibérico*; paella (especially in its birthplace, Valencia); Manchego cheese
Drink: *vino tinto* (red wine) from La Rioja wine region; *vino blanco* (white wine) from Galicia; or *fino* (sherry) from Jerez de la Frontera

In a word
¿Que pasa? (What's happening?)

Celebrate this
Somewhere in Spain a fiesta is taking place: from messy La Tomatina to religious Semana Santa (Easter), from San Juan bonfires to Las Fallas' fireworks, there's a celebration here for every occasion.

Random fact
Sobrino de Botín restaurant in the centre of Madrid dates to 1725 and is recognised as the world's oldest.

1. Bar-hopping in Barcelona is an essential pastime for visitors and locals alike

2. The historic fortified hilltop of Dalt Vila in Ibiza Town was first settled by the Phoenicans, but Ibiza harbour now sees yachts and sightseers

3. Spain produces some of the tastiest hams, sausages and olive oils you'll find in Europe

4 The Picos de Europa mountains in northwest Spain, straddling Asturias, Cantabria and Castilla y León, are a beguiling landscape to explore on foot or bicycle

CAPITAL MADRID // **POPULATION** 50 MILLION // **AREA** 505,370 SQ KM // **OFFICIAL LANGUAGE** SPANISH (SEVERAL REGIONAL LANGUAGES OFFICIAL LOCALLY)

Spain

Spain is a heady mix of curious traditions and a relentless energy that propels Spaniards into the future. You see it in the architecture: the Islamic confections of Andalucía and soaring Gothic cathedrals share the country with avant-garde creations by Gaudí and Santiago Calatrava. You taste it in the food: you're just as likely to find three generations of the same family in the kitchen as you are chefs with three Michelin stars honouring the innovations of new Spanish cuisine. There are jagged sierras, wild coastlines, soul-stirring flamenco and world-class art galleries with collections spanning the centuries. But for all this talk of past and future, Spaniards live very much in the present, celebrating life and their country in a seemingly nonstop round of fiestas.

S CAPITAL COLOMBO // POPULATION 22.9 MILLION // AREA 65,610 SQ KM // OFFICIAL LANGUAGES SINHALA & TAMIL

Sri Lanka

Buffeted by the spice-scented trade winds of the Indian Ocean, but scarred by civil war, Sri Lanka is slowly rebuilding its reputation as the southern gateway to South Asia. While investigations continue into the bloody final years of conflict between Buddhist Sinhalese and Hindu Tamils, Sri Lankans are looking to a new future of peace and prosperity, with returning tourists playing a major role. Even the terrorist attacks of April 2019 failed to deter visitors from Sri Lanka's beaches, surf breaks, national parks and temples for long. Those who come find endless sweeps of golden sand, grasslands teeming with leopards and wild elephants, tea plantations painting the hills in emerald swirls and the towering ruins of ancient Buddhist kingdoms, all easily accessible from the resorts and colonial-era cities dotted around the shoreline.

Best time to visit
December to March, to avoid the southwest monsoon

Top things to see
- Museums, monuments and cafe culture in the frenetic capital, Colombo
- Ruined palaces and super-sized Buddhas in the old royal capital of Polonnaruwa
- A garden of stone columns and towering *dagobas* (stupas) in Anuradhapura
- Exquisite frescoes and epic views in the ancient fortress at Sigiriya
- A very different side of Sri Lanka in the Tamil city of Jaffna

Top things to do
- Kick back on the sparkling sands of Sri Lanka's southern beaches
- Climb sacred Adam's Peak to watch the rising sun cast its rays over the island
- Watch elephants congregate at 'the Gathering' in Minneriya National Park
- Surf the famous breaks at Arugam Bay or Weligama
- Make the pilgrimage to the Temple of the Tooth in World Heritage–listed Kandy

Getting under the skin
Read: Shyam Selvadurai's unconventional love story, *Funny Boy*; Ashok Ferrey's novel of returning diaspora, *Serendipity*; or Samanth Subramanian's war exposé, *This Divided Island*
Listen: to *baila* (Sri Lankan dance music); or the quirky metal of Stigmata and Funeral in Heaven
Watch: Purahanda Kaluwara's haunting family drama, *Death on a Full Moon Day*; or political biopic *Ginnen Upan Seethala* (The Frozen Fire) by Anuruddha Jayasinghe
Eat: 'hoppers' (or more properly *appa*), pancakes made from fermented rice and coconut milk
Drink: toddy (a local wine made from fermented palm sap); or *arrack*, the same thing but distilled and bottled

In a word
Ayubowan (May you live long)

Celebrate this
With the treatment of Sri Lanka's elephants in the spotlight, campaigners such as the Otara Foundation are pressuring the government to update the country's century-old animal rights laws; a new Animal Welfare Bill is currently awaiting approval by parliament.

Random fact
Sri Lanka gave the world cinnamon – the island has been trading the spice since at least 2000 BCE.

1. The central highlands of Sri Lanka are the perfect place to grow tea plants; different varieties prefer different elevations and soils so there is a great variety to explore

2. Take the train to the misty city of Kandy where the Temple of the Sacred Tooth Relic houses a tooth of the Buddha

3. Tangalle and the rest of Sri Lanka's south coast has some fantastic beaches and whale-watching tours depart regularly

MATT MUNRO | LONELY PLANET | MATT MUNRO | LONELY PLANET // ANTON PETRUS | GETTY IMAGES

1. A sandstorm, known as a *haboob,* submerges Khartoum in sand

2. The pyramids of Meroe are the remarkable remnants of a Nubian kingdom from 2500 years ago, with hundreds of royal cemeteries

3. Food is in short supply in strife-torn Darfur so a woman soaks mokhat berries for three days before they're able to be consumed

S CAPITAL KHARTOUM // POPULATION 45.6 MILLION // AREA 1,861,484 SQ KM // OFFICIAL LANGUAGES ARABIC & ENGLISH

Sudan

Once home to the ancient Kingdom of Kush, Sudan was a place famed for its gold, ivory and incense. While the march of time (and of the Sahara) has changed the landscape vastly, impressive relics of its past remain. Look no further than the pyramids rising from the sands of the northeast, which outnumber the total found in Egypt, to understand the power of its black pharaohs. Yet not all has succumbed – dive beneath its Red Sea waters and you'll find aquatic riches along its reefs. Today, Sudan's people also remain as strong as they are famously hospitable; having endured years of civil war in the 21st century, they collectively stood up to remove their autocratic military president of 30 years in 2019.

Best time to visit
October to March (the dry season)

Top things to see
- Sands enveloping the pyramids at the royal cemetery of Begrawiya, the resting place of Kush's Meroitic Pharaohs
- The Egyptian temple of Soleb, built in the 14th century BCE by Amenhotep III
- Khartoum's National Museum, the best in Sudan
- The two Kushite temples of Naqa, both examples of the ancient kingdom's architecture

Top things to do
- Let your eyes wander in the souqs of Kassala
- Explore the holy mountain of Jebel Barkal and the vestiges of the Temple of Amun
- Survey the 3500-year-old remains of the Kingdom of Kush from atop Kerma's mudbrick fortress
- Witness fluid dynamics in action as the two Niles, Blue and White, meet and meld in Khartoum

Getting under the skin
Read: *The Grub Hunter* by Amir Tag Elsir, the story of a former security agent turned novelist
Listen: to Amira Kheir's *Mystic Dance*, an enthralling mix of modern music rooted in tradition
Watch: *Talking About Trees*, a documentary about four directors that sheds light on the nation's recent troubles; *The Devil Came on Horseback*, a powerful documentary exposing the war crimes in Darfur
Eat: *fuul* (stewed brown beans) for breakfast, complete with cheese, egg, salad and flatbread
Drink: sweet black *shai* (tea)

In a word
Salaam aleikum (Peace upon you)

Celebrate this
Fearless Alaa Salah: the student who stood atop a car outside President Al Bashir's residence and led chants for peaceful revolution in 2019.

Random fact
Pyramids in the Kingdom of Kush weren't just reserved as resting places for members of the royal family but also for priests and high-ranking officials.

OMER KOCLAR | SHUTTERSTOCK // DENIS BURDIN | SHUTTERSTOCK // LIBA TAYLOR | GETTY IMAGES

1. The giant freshwater otter grows up to 1.7m in length and thrives in Suriname's waterways

2. Approaching the village of Palumeu, upstream along the Boven Tapanahoni River

3. Discover Dutch colonial architecture in Paramaribo; the Dutch invaded in 1667 and plantations used enslaved Africans to produce sugar, cocoa, coffee and cotton

S **CAPITAL** PARAMARIBO // **POPULATION** 609,569 // **AREA** 163,820 SQ KM // **OFFICIAL LANGUAGE** DUTCH

Suriname

One of South America's smallest countries, Suriname packs in a surprising jumble of cultures. The heavily forested nation is home to a mix of people descended from enslaved West Africans; Javanese, Chinese and Indian labourers; indigenous groups; and Dutch, Lebanese and Jewish settlers. Paramaribo, where half the population resides, is a blend of synagogues and mosques, Indian roti shops and Chinese dumpling houses spread among Dutch colonial buildings. Outside the capital, dirt tracks and meandering rivers lead to Suriname's natural wonders, which include savannahs, vast swathes of protected rainforest and remote beaches that are a major breeding site for endangered sea turtles.

Best time to visit
February to April and August to November

Top things to see
• Paramaribo, Suriname's vivacious capital and a Unesco World Heritage Site, with its fantastic blend of cultures (and cuisines)
• The Central Suriname Nature Reserve, a vast reserve of diverse ecosystems and wildlife
• The indigenous village of Palumeu, on the remote banks of the Boven Tapanahoni River

Top things to do
• Take a boat journey along the Upper Suriname River stopping off at Maroon villages
• Watch over the sands of Galibi Nature Reserve as giant leatherback turtles emerge from the sea to lay their eggs on the beach (April to July)
• Spy howler monkeys in the jungle canopy of the Brownsberg Nature Reserve

Getting under the skin
Read: *Tales of a Shaman's Apprentice* by Mark Plotkin, about an ethnobotanist's discoveries living among Amerindians in Suriname and Brazil
Listen: to flautist Ronald Snijders' mix of jazz, beatboxing, classical melodies and indigenous sounds
Watch: *Let Each One Go Where He May*, Ben Russell's hypnotic account of two brothers travelling from the city to the forests of the interior
Eat: *bami goreng* (fried noodles); or *pom* (a casserole of chicken and elephant ear root)
Drink: ice-cold Parbo beer

In a word
Fa waka? (How are you?)

Celebrate this
The country hasn't qualified for a football World Cup, but Surinamese footballers representing the Netherlands – including Ruud Gullit and Clarence Seedorf – are among the greatest players ever.

Random fact
Although Dutch is the official language, the lingua franca is Sranan Tongo, an English-based creole with African, Portuguese and Dutch roots.

DAVID HAVEL | SHUTTERSTOCK // RENÉ HOLTSLAG | SHUTTERSTOCK // IMAGEBROKER | AWL IMAGES

S CAPITAL STOCKHOLM // POPULATION 10.2 MILLION // AREA 450,295 SQ KM // OFFICIAL LANGUAGE SWEDISH

Sweden

Sweden has long punched above its weight. First it was Vikings, then a northern Europe-wide empire, then the contagious Scandi-pop of ABBA, then Ericsson phones and Ikea's flat-pack revolution, then Sweden's darker side, brought to life in *Wallander* and *The Girl with the Dragon Tattoo*. Indeed, the real-life locations for Sweden's silver-screen outings have become tourist attractions in their own right, alongside more established sights such as graceful Stockholm and the scenic Bohuslän coast. The cities are changing fast as Sweden manages an unprecedented influx of refugees that is changing the nation's own perception of itself. But away from the urban centres are the scattered islands, forests full of berries, villages of red weatherboard houses and fresh air activities that make this a playground for nature lovers.

Best time to visit
May to August for sunshine, December to March for skiing

Top things to see
- The glittering beauty of Stockholm from Söder Heights
- A manageable portion of the 24,000 islands of the Stockholm archipelago, ideally on a boat cruise
- The midnight sun above the Arctic Circle in Abisko National Park, deep in the heartland of the Sami people
- Glass blown to perfection in Glasriket (the Glass Kingdom)
- Rocky islands and idyllic fishing villages on the picturesque Bohuslän coast

Top things to do
- Educate yourself in the museums of Uppsala, home of Sweden's oldest university
- Throw yourself into the great outdoors cycling around Gotland, or cross-country skiing and hiking in frozen Norrland
- Bed down for a night in the supercool Icehotel at Jukkasjärvi
- Dig into history in Stockholm's Skansen open-air museum and maritime Vasamuseet
- Hang out with verified Vikings at the Foteviken Viking Reserve

Getting under the skin
Read: crime-busting novels courtesy of Stieg Larsson's blockbuster Millennium Trilogy or Henning Mankell's Kurt Wallander series; or the sweet tales of *Pippi Longstocking*
Listen: to ABBA, Roxette and Robyn for world-conquering pop music; or The Hives for Swedish rock
Watch: any of the dozens of films directed by the great Ingmar Bergman; the acclaimed *My Life as a Dog*; or the sublime, subtle vampire film *Let the Right One In*
Eat: *lax* (salmon) in its various guises; game such as elk and reindeer; and the requisite *köttbullar* (Swedish meatballs), served with mashed potatoes and lingonberry sauce

Drink: *kaffe* (coffee); Absolut vodka; or the beloved *aquavit* and *öl* (beer)

In a word
Jättebra! (Fantastic!)

Celebrate this
From protesting outside Sweden's parliament to global activist, Greta Thunberg's rise as a spokesperson for environmental issues has been rapid and well publicised, helping raise green awareness like nobody else of her generation.

Random fact
The best-known Swedish inventor is Alfred Nobel, who discovered dynamite and also, ironically, gave money to found the Nobel Institute, giver of peace prizes.

1. The Northern Lights ripple above Swedish Lapland; the Aurora Sky Station in Abisko National Park is a great place to view them from August to April

2. Central Sweden is filled with lakes and forests

3. Södermalm is Stockholm's creative engine room, with indie fashion boutiques, vintage stores, galleries, bars, espresso labs and music venues

S

Best time to visit
Year-round

Top things to see
- The majestic Matterhorn dominating the chic ski resort town of Zermatt
- Lake Geneva: whether you paddleboard on it or dine alfresco beside it
- Europe's largest glacier, the Aletsch, sticking out its iced serpentine tongue for 23km
- The might and power of the thunderous Rheinfall, best admired from the Känzeli viewing platform in Schloss Laufen
- The medieval chateau-village of Gruyères and its working cheese dairy

Top things to do
- Shop for urban fashion and Switzerland's best chocolate in hip Zürich
- Wallow, quite literally, in modern architecture at the spa of Therme Vals
- Hike through flower-dotted meadows and larch woodlands past shimmering blue lakes and rocky outcrops in the Swiss National Park
- Check out live jazz lakeside at the legendary Montreux Jazz Festival in July
- Take a train trip through some of the most magnificent Alpine scenery in the world

Getting under the skin
Read: *At Home*, a collection of short stories by Zürich cabaret artist Franz Hohler
Listen: to folk band Sonalp, a fusion of Swiss yodelling, cow bells and world music; or Bern-born singer-songwriter Sophie Hunger
Watch: the James Bond 1960s classic *On Her Majesty's Secret Service* for action-packed shots of Bern, Grindelwald and snowy Saas Fee
Eat: *rösti* (a shredded, crispy potato bake) and *würste* (sausages) in German-speaking Switzerland, and a cheesy fondue or raclette when there's snow
Drink: prestigious Calamin or Dézaley *grand cru* from the Unesco-protected vineyards of Lavaux facing Lake Geneva

In a word
Grüezi (Hello in Swiss German)

Celebrate this
In 1957 Max Miedinger can't have imagined that his new typeface would be the world's most popular over six decades later, but such a hit was Helvetica (named after Switzerland's pre-Roman inhabitants) that it's still online and in print everywhere.

Random fact
Müsli (muesli) was invented in Switzerland at the end of the 19th century; the most common form is the ever-popular Birchermuesli.

1. A *Tschäggä* prepares to terrify the people of the Lötschental Valley in the Swiss Alps as a carnival tradition

2. Swiss St Bernard dogs awaiting an SOS call in Champex-Lac

3. The Limmat River flows through Zurich and into Lake Zurich

JONATHAN GREGSON | LONELY PLANET // MATT MUNRO | LONELY PLANET // JUSTIN FOULKES | LONELY PLANET

S   **CAPITAL** BERN // **POPULATION** 8.4 MILLION // **AREA** 41,277 SQ KM // **OFFICIAL LANGUAGES** GERMAN, FRENCH, ITALIAN & ROMANSCH

Switzerland

Few places invite self-indulgence quite as much as Switzerland. Whether you're into Alpine action, cool urban cities, lakeside pursuits or hiking through green fields, you'll find a way to treat yourself here. Roam its charming castles and villages, soak in mountain spa waters, pause for a schnapps or hot chocolate in a chalet strewn with geranium boxes, admire the views on a mountain train between pine wood and glacial peak, or glide silently on skis around the pyramid-shaped Matterhorn. This small, landlocked country of four languages was an essential stop on every 18th-century Grand Tour, it was the place where winter tourism was born and where Golden Age mountaineers scaled new heights, and it has continued to captivate travellers ever since.

S   **CAPITAL** DAMASCUS // **POPULATION** 19.3 MILLION // **AREA** 187,437 SQ KM // **OFFICIAL LANGUAGE** ARABIC

Syria

It's impossible to imagine what the future holds for Syria. Approximately 5.5 million Syrians have left the country as refugees while over six million are internally displaced due to the brutal conflict which continues to rage, now mostly confined to Syria's northwest. It's cold comfort to say that this once-glorious civilisation has survived the horrors of war many times before throughout the ages. Syria has always stood at history's crossroads - its Roman ruins, Crusader castles and ancient cities are testament to that, though many of its historic sites have suffered severe, irreparable damage.

Best time
March to May and September to November

Places of interest
- Damascus' old city, with its wriggling maze of cobblestone lanes
- Crac des Chevaliers, the hilltop-perched Crusader castle
- The Umayyad Mosque in Damascus for its mosaic and marble internal courtyard
- Bosra's 15,000-seat Roman theatre and citadel
- The National Museum, where a vast collection of Syria's 11 millennia of heritage has been protected

Local activities
- The climb up to Qalaat Salah ad-Din's ridgetop perch provides vistas across the coastal plain
- The old souqs of Damascus are crowded with locals haggling for spice, textiles and household goods
- Birj Islam beach, north of Lattakia, is a popular swimming spot in summer
- Aleppo's citadel, remarkably intact despite the old city district's destruction, is where people come for sunset views
- Cafe-life revolves around socialising while smoking fruit-flavoured tobacco through a *nargileh* (water-pipe)

Getting under the skin
Read: *No Knives in the Kitchens of this City* by Khalid Khalifa; and *No Turning Back* by Rania Abouzeid
Listen: to Lena Chamamian's *Shamat*
Watch: Critically acclaimed documentaries *For Sama*, directed by Waad Al-Kateab, and *The Cave*, directed by Feras Fayyad
Eat: *booza*, an elastic, melt-resistant ice cream
Drink: *shay na'ana* (mint tea)

In a word
Ahlan wa sahlan (Welcome)

Celebrate this
Damascus and Aleppo are among the oldest continuously inhabited cities on Earth and Ugarit is home to one of the world's first alphabets.

Random fact
One out of every three Syrians is under 15 years.

1. A Syrian plays the *mizmar*, a wind instrument, to welcome back a victorious women's football team in Amuda

2. The view from the Mohammadieh Noria (water wheel) along the Orontes river in the city of Hama; the water wheels await Unesco protection

3. Fakhr-al-Din al-Ma'ani castle overlooks the ruins of the great city of Palmyra, a cultural crossroads from the 1st century

S

T CAPITAL TAIPEI // POPULATION 23.8 MILLION // AREA 35,980 SQ KM // OFFICIAL LANGUAGE MANDARIN

Taiwan

One of the four Asian tigers, the island of Taiwan is divided from the Chinese mainland by the Taiwan Strait – and by a yawning political divide. Politically known as the Republic of China (ROC), Taiwan has been ruled by the nationalist government founded by Chiang Kai-shek since 1945, although the island is nominally claimed by the People's Republic of China (PRC). Politics aside, this is a land where skyscrapers rub shoulders with misty mountains and rugged coastlines. Away from the futuristic capital, Taipei, Taiwan is a tapestry of forested peaks, giant Buddhas, hot springs, basalt islands and tribal villages. No wonder the island was christened Ilha Formosa (Beautiful Island) by Portuguese sailors when they passed by in the 1540s.

Best time to visit
September to November, for ravishing autumn colours, and January or February for the annual lantern festival

Top things to see
- The view over Taipei from the 509m-high Taipei 101 tower
- A staggering array of Chinese artefacts in Taipei's National Palace Museum
- Historic temples and gourmet dining in Tainan, the ancient capital
- Stunning roadside scenery while cycling the South Cross-Island Hwy or Hwy 11
- Basalt outcrops and delightfully sculptural stone fish traps on the Penghu islands

Top things to do
- Trek through marble canyons beside a jade-green river in Taroko Gorge
- Learn to drink tea the Taiwanese way in the tea gardens of Pinglin
- Hike to hidden waterfalls and hot springs in the jungles around Wulai
- Learn about Taiwan's aboriginal heritage among the Yami tribal people on lovely Lanyu island
- Test your commitment to adventurous eating at one of Taipei's legendary night markets

Getting under the skin
Read: Hsiao Li-hung's classic love story *A Thousand Moons on a Thousand Rivers*
Listen: to the increasingly diverse Taiwanese music scene including Sun Sheng Xi, Soft Lipa and No Party for Cao Dong
Watch: *Seediq Bale*, an epic film about the Wushe Rebellion; and Ang Lee's *Eat Drink Man Woman* for an insight into modern Taiwanese culture
Eat: *chòu dòufu* ('stinky tofu'), marinated in a brine made from decomposing vegetables and shrimps
Drink: Ali Shan tea, a delicate oolong grown at high altitude in Taiwan's central mountains

In a word
Chīfàn le ma? (Have you eaten yet?)

Celebrate this
With a comprehensive and successful response grounded in science, community cooperation and excellent health infrastructure, Taiwan's approach to managing Covid-19 and protecting its economy was one of the world's best.

Random fact
Both instant noodles and bubble tea (also known as 'boba' or 'pearl milk' tea) were invented in Taiwan.

1. Liberty Square in Taipei dates from the 1970s

2. Cihmu Bridge is at the upper end of Taroko Gorge; the landscape of Taroko National Park climbs from sea level to 3400m in a short distance

3. Must-try Taiwanese dishes include beef noodle soup, stinky tofu and *xiao long bao* (soup dumplings)

4 The view of Taipei 101 from Elephant Mountain

(T) CAPITAL DUSHANBE // POPULATION 8.9 MILLION // AREA 143,100 SQ KM // OFFICIAL LANGUAGE TAJIK

Tajikistan

A Persian speaker in a Turkic world, tucked in a mountain cul-de-sac at the furthest corner of the former USSR, Tajikistan is a powerhouse adventure-travel destination whose world-class potential remains ripe for exploration. Aside from bustling Silk Road towns and colourful bazaars, Tajikistan's main pull is the Pamir Mountains, a high-altitude plateau of intensely blue lakes, Kyrgyz yurts and rolling valleys that has impressed everyone from Marco Polo to Francis Younghusband. If your favourite places include Tibet, Bolivia or northern Pakistan, chances are you'll be blown away by little-known Tajikistan.

Best time to visit
June to September (mountains), March to May, October to November (Dushanbe and lowlands)

Top things to see
- Pamir Hwy, one of the world's great road trips, linking Khorog and Osh over the 'Roof of the World'
- The Wakhan Corridor, a gloriously scenic valley of Silk Road forts, Buddhist stupas and 7000m peaks on the remote border with Afghanistan
- Marguzor Lakes, a string of seven turquoise lakes and homestays near Penjikent
- Istaravshan, a historic town with a great bazaar, an ancient citadel and mosques and madrasas hidden in its side streets

Top things to do
- Fasten your seatbelt on the scenically outrageous flight from Khorog to Dushanbe, the only route in the former Soviet Union where pilots were awarded hazard pay
- Trek the Fan or Pamir Mountains, up there with the world's best mountain scenery
- Stay overnight in a mountain homestay or yurtstay to experience the region's humbling hospitality
- Stroll the neoclassical facades of central Dushanbe, pausing to admire the world's largest teahouse and second tallest flagpole.

Getting under the skin
Read: *Land Beyond the River* by Monica Whitlock, for a rundown of recent history from the BBC correspondent
Listen: to a folk singer belting out a Rudaki poem to the tune of a six-stringed *rubab* (Persian lute) in the western Pamir
Watch: Bakhtyar Khudojnazarov's road movie *Luna Papa*; or Jamshed Usmonov's black comedy *Angel on the Right*
Eat: a communal bowl of Pamiri *kurtob* (a deliciously cool and creamy mix of flat bread, yoghurt, onion, tomatoes, chives and coriander)
Drink: a locally brewed Sim-Sim *piva* (beer) beside the fountains of Dushanbe's Ayni opera house

In a word
Roh-i safed (Have a good trip)

Celebrate this
An improvement in relations with neighbouring Uzbekistan made life easier for many Tajiks in 2018, with the reintroduction of border crossings, visa-free visits and the first flights between Dushanbe and Tashkent for 20 years.

Random fact
Ancient Sogdian, the language of the Silk Road, is still spoken in parts of Tajikistan's Yagnob Valley.

1. Alaudin Lakes in the Chapdara River valley are high in the Fan Mountains
2. A game of *buzkashi* being played in Hisor; the teams on horseback attempt to place a goat's carcass in the opposing team's goal
3. The earliest incarnations of Hisor fort were built 3000 years ago
4 Climbers above 6000m on Korzhenevskaya peak in Tajik National Park; there are many unclimbed mountains in the Pamirs

(T) **CAPITAL** DODOMA // **POPULATION** 58.6 MILLION // **AREA** 947,300 SQ KM // **OFFICIAL LANGUAGES** SWAHILI & ENGLISH

Tanzania

Some countries make noise. Tanzania makes music. Its largest soundstage – the vast Serengeti plains – hosts earth's most spectacular natural show, the great migration. Not only do grasses sway to the millions of wildebeest hoof beats, but so do the hearts of lucky visitors who drop in each year. Melodies of different sorts, whether the chants of leaping Maasai warriors, the lilting songs of Kilimanjaro guides or the rhythmic lapping of turquoise waters on Zanzibar's shores, are as intrinsic as Tanzania's epic visuals – dramatic Rift Valley landscapes, colourfully clad locals and spellbinding wildlife. Although the nation is home to one of Africa's most diverse populations, tribal rivalries are almost non-existent. As you'd expect from a great maestro, harmony reigns.

Best time to visit
June to October or December to January

Top things to see
- An unparalleled wealth of African wildlife encircled by volcanic walls, as viewed from the top of Ngorongoro Crater's jagged rim
- Dawn breaking over the African savannah from your spot atop Kilimanjaro
- The lively and colourful local markets in the Usambara Mountains
- Swollen herds of elephants and giant baobabs along the riverbanks of Ruaha National Park
- The interior of Kilwa Kisiwani's Great Mosque lit by a sun low on the horizon

Top things to do
- Follow a stream of countless wildebeests and zebras as they migrate across the great Serengeti plains (with predators in tow)
- Find Swahili surprise after Swahili surprise while wandering the depths of Stone Town on Zanizbar
- Drift or wall dive along the steeply dropping shelf off the island of Pemba
- After chimpanzee tracking, chill on a tropical beach in Mahale Mountains National Park
- Take to the Rufiji River on a boat safari within the vast Selous Game Reserve

Getting under the skin
Read: *The Worlds of a Maasai Warrior* by Tepilit Ole Saitoti, a coming-of-age story from the plains of the Serengeti
Listen: to Vanessa Mdee, for her mix of R&B and Bongo Flava (a form of Swahili hip hop incorporating Afrobeat and arabesque melodies)
Watch: *Shadow Boy*, a short film that follows a Zanzibari's efforts to save a life; *Kilimanjaro – To the Roof of Africa*, a stunning IMAX documentary by David Breashears
Eat: *ugali* (maize and/or cassava flour porridge); or *mishikaki* (marinated meat kebabs)
Drink: *mbege* (banana beer); or *uraka* (a brew made from cashews)

In a word
Hakuna matata (No worries)

Celebrate this
Barbara Gonzalez: at the age of 30 she became the CEO of Simba, one of Africa's top football teams.

Random fact
The Anglo-Zanzibar conflict on 27 August 1896 lasted a grand total of 38 minutes.

1. A Maasai herder leads his cattle to water
2. Each year vast herds of wildebeest migrate to their winter grazing grounds in the southern Serengeti
3. Dhows are traditional sailing boats on Zanzibar, the semi-independent island group about 23 miles from Tanzania

T

T // CAPITAL BANGKOK // POPULATION 69 MILLION // AREA 513,120 SQ KM // OFFICIAL LANGUAGE THAI

Thailand

For millions of travellers, Thailand is the doorway to Southeast Asia: a first taste of the region's magical mix of history, religion, culture and natural wonders. From the temple spires of Bangkok to the white beaches of the southern islands, this is Asia as most people have always imagined it: dramatic, tropical and blessed by one of the most tantalising cuisines on the planet. Backpackers and families from across the globe throng to Thailand's beaches and jungles, temple towns and tribal villages, trekking its rainforest trails and diving on its coral reefs. Whatever piques your interest, you'll find a version of Thailand that satisfies it, be that strolls around Buddhist stupas, or full-moon parties.

Best time to visit
November to April, to escape the main rainy season

Top things to see
- Culture, chaos, cacophony and Buddhist culture in Bangkok, Thailand's energetic capital
- The atmospheric ruins of the ancient Thai capitals at Sukhothai and Ayuthaya
- Pristine rainforests teeming with wildlife at Khao Yai and Khao Sok National Parks
- Moving relics from the Thailand–Burma Railway at Kanchanaburi
- Idyllic islands, from party-ready Phi-Phi to the castaway isles of Ko Tarutao National Marine Park

Top things to do
- Learn to conjure up a Thai feast on a cooking course in Bangkok, Phuket or Chiang Mai
- Test your head for heights by climbing the awesome karst outcrops at Krabi
- Discover tribal culture, jungle fauna, and natural hot springs in the mountains around Chiang Rai and Chiang Mai
- Dive with magnificent megafauna off Ko Tao or around the Similan Islands
- Dance barefoot on the sands at a full-moon bash on Ko Pha-Ngan

Getting under the skin
Read: *Monsoon Country* by Nobel Prize–nominated Thai author Pira Sudham
Listen: to the nostalgic rock of Carabao; or the jazz compositions of His Majesty King Bhumibol Adulyadej
Watch: Chatrichalerm Yukol's historical romp *Legend of Suriyothai*; or the unsettling folk-horror of Sitisiri Mongkolsiri's *Inhuman Kiss*
Eat: *tom yam kung* (hot and sour prawn and lemongrass soup); or *pàt tai*, Thailand's favourite fried noodles
Drink: Singha and Chang beer; or potent Sang Som whisky, served with Red Bull and ice

In a word
Sanuk (Fun) – the cornerstone of the Thai psyche

Celebrate this
As a campaigner, Human Rights Commissioner and senator, Tuenjai Deetes has spent a lifetime supporting Thailand's tribal communities, helping 90,000 people secure Thai nationality – and access to medical services, education and employment.

Random fact
Every Thai house has a *san phra phum* (spirit house) for the animist spirits dwelling on the site to live in.

1. The spectacular limestone karst islands of Ao Phang Nga National Park in southern Thailand
2. Songkran is the Thai new year celebration every April and an excuse for a massive water fight
3. Sukhothai Historical Park includes the remains of 21 historical sites and four ponds within the old walls
4. Pai night market in Mae Hong Son is a sensory extravaganza with food from northwest Thailand

T

1. Colourful corals support delicate ecosystems in the seas around Dili

2. The waterfront is the focus of activity in capital city Dili

3. Women wearing traditional dress during the Celebration of Independence

(T) **CAPITAL** DILI // **POPULATION** 1.3 MILLION // **AREA** 14,874 SQ KM // **OFFICIAL LANGUAGES** PORTUGUESE & TETUN

Timor-Leste

Becoming the first sovereign state of the 21st century was no small victory for Timor-Leste, which fought hard for its independence from a 24-year occupation by Indonesian forces. While the first years of freedom were tumultuous, this impoverished former Portuguese colony is steadily rising to its feet. Stunning natural wonders, from mist-shrouded mountain peaks to rugged, deserted beaches, await those ready to challenge the lack of tourism infrastructure and diabolical – but improving – roads. Outside Dili, many people live in traditional dwellings as they have for centuries. Pristine dive sites are still being discovered. Everywhere you go, you'll meet a proud people curious to find out what brings you to this corner of the world.

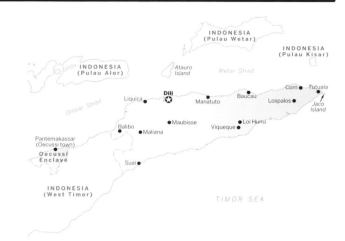

Best time to visit
It's hot all year, but May to November is driest

Top things to see
- Atauro Island, a starkly beautiful isle off Dili that beckons divers, hikers and beach bums
- Traditional architecture, including the iconic stilted Fataluku houses in the nation's east
- The sun setting at the 300-year-old Balibo Fort
- Local life on the beaches of Dili, from spirited soccer matches to fishers hauling in their loads

Top things to do
- Strap on a tank and explore the excellent dive sites that dot the nation's north coast
- Learn about Timor-Leste's tumultuous history at Dili's excellent Resistance Museum
- Spend an afternoon lazing on the powder-white sands of uninhabited, sacred Jaco Island
- Hike to the summit of Mt Ramelau, the nation's highest peak, to watch dawn break over Timor

Getting under the skin
Read: Gordon Peake's *Beloved Land: Stories, Struggles and Secrets from Timor-Leste*, an affectionate look at the country's characters and the challenges of creating a brand-new nation
Listen: to *tebe*, festive folk music
Watch: *Balibo*, the confronting feature film about a group of Australia-based journalists executed while reporting on Indonesia's activities in Timor in 1975
Eat: 'fish on a stick' served at coastal roadhouses
Drink: Arabica coffee grown in the hills above Dili

In a word
Bom dia! (Good morning!)

Celebrate this
Xanana Gusmão, nicknamed the 'Nelson Mandela of Southeast Asia', was Timor-Leste's first president. This Timorese-of-all-trades is also renowned as a freedom fighter, human-rights champion and published poet.

Random fact
A small patch of Timor-Leste (the Oecusse enclave) sits about 80km from the rest of the country.

T

Best time to visit
Mid-July to mid-September

Top things to see
- The waters of Lac Togo from a seat in a traditional pirogue
- Buffaloes, ostriches and antelope on the grasslands of Parc Sarakawa
- Unesco World Heritage–listed area, Koutammakou, famous for its fortified clay-and-straw *tata* compounds
- The crossroads town of Dapaong, which has a stunning cave settlement nearby

Top things to do
- Soak up the mellow vibes, live music and colourful markets of Lomé
- Unwind on the sands of Coco Beach, or walk the boardwalk to a drink at a seafront bar
- Hike through the cocoa and coffee plantations around Kpalimé before cooling off in a waterfall
- Visit the vibrant Vogan Friday Market northwest of Aného
- Journey by pirogue to Togoville, the former seat of the Mlapa dynasty, then immerse yourself in voodoo traditions

Getting under the skin
Read: *Do They Hear You When You Cry* by Fauziya Kassindja, the story of a young girl's life turning with the death of her father and impending female genital mutilation
Listen: to King Mensah, 'the Golden Voice of Togo'; or Toofan, for the best of modern Togolese beats
Watch: *The Blooms of Banjeli* – shot in 1914, but not released until 1986, it documents life in a small village
Eat: *pâte* (a dough-like substance made of corn, manioc or yam) accompanied by sauces such as *arachide* (peanut and sesame) or *gombo* (okra)
Drink: *tchoukoutou* (fermented millet), the preferred tipple in the north

In a word
Be ja un sema (How are you? in Kabyé)

Celebrate this
Claude Grunitzky, a journalist, editor and entrepreneur who champions young African's voices through the media platform True Africa: he also founded the fashion and music publication Trace, and Trace TV.

Random fact
Togo is one of only two countries in Africa where more than 40% of the land is suitable for agriculture.

1. A traditional tower house in the Koutammakou area

2. The Independence Monument in Lomé marks Togo's independence from France in April 1960

3. Togo has just 56km of coastline but makes the most of it with the capital city Lomé located at the west end, where a fishing fleet is based

CAPITAL LOMÉ // POPULATION 8.6 MILLION // AREA 56,785 SQ KM // OFFICIAL LANGUAGE FRENCH

Togo

Although just a tad larger than Tanzania's Selous Game Reserve, Togo proves size has little to do with variety. Its landscapes range from palm-fringed Atlantic beaches and lagoons in the south to lakes and forested hills at its centre, and in the north, open savannah plateaus dotted with African wildlife. Its human geography, too, boasts a diversity way out of proportion to the country's size: a staggering 40 ethnic groups live within its borders. The fortified compounds of Koutammakou, however, are a historic reminder that the country's many peoples have not always got along. Intermittent unrest has also existed since independence, and due in no small part to almost six decades of rule by the same family, it is growing.

(T) **CAPITAL** NUKU'ALOFA // **POPULATION** 106,095 // **AREA** 747 SQ KM // **OFFICIAL LANGUAGES** TONGAN & ENGLISH

Tonga

Say goodbye to tourist hype – the Kingdom of Tonga is pure Polynesia. You won't find resorts selling packaged fun and you won't have to look too hard for authentic experiences because they're everywhere. From the monarchy and Christian church services to feasts and traditional dancing, Tonga pulsates with cultural identity. The backdrop to this vibrant way of life is sublime: pristine beaches, rainforests, soaring cliffs and underwater caves. Go with the flow and keep it slow on Tonga time.

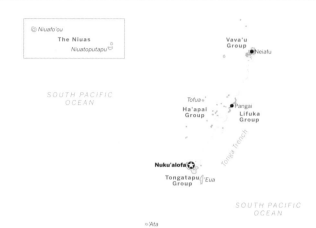

Best time to visit
May to October when it's cooler and drier

Top things to see
• The 13th-century Ha'amonga 'a Maui trilithon – the 'Stonehenge of the South Pacific'
• Tongan arts, crafts and pyramids of tropical fruit at the Talamahu Market
• Fire-dancing inside Hina Cave while feasting on Tongan fare
• Hundreds of Mapu'a 'a Vaea blowholes spurting at once along a 5km stretch of coastline
• A colourful Sunday church service with magnificent, booming choirs

Top things to do
• At a responsible distance, view humpback whales that come to breed in Tongan waters between July and November
• Kayak around Vava'u's turquoise waterways and islands to deserted beaches
• Explore the remote doughnut-shaped volcanic isle of Niuafo'ou
• Hike through laid-back 'Eua's limestone caves and tropical forests to dramatic cliff edges
• Laze on Uoleva's spectacular beach and, if you're lucky, watch whales breaching offshore

Getting under the skin
Read: Epeli Hau'ofa's slim but hilarious *Tales of the Tikongs*, which casts a satirical eye over island life
Listen: to *Dance Music of Tonga* by Mālie! Beautiful!
Watch: *My Lost Kainga*, about a Tongan woman raised in Australia who returns to the islands to discover her culture
Eat: a feast prepared in an *umu* (earthen oven): taro and yam, roast suckling pig, coconut, fish and shellfish
Drink: kava, the murky tipple that's a part of the Tongan experience

In a word
Ha'u 'o kai (Come and eat!)

Celebrate this
The Tongan traits of resilience, resourcefulness and faith are behind the astonishing true story – soon to be made into a movie – of six Tongan teenagers shipwrecked and stranded on a deserted island for 15 months in the 1960s.

Random fact
From midnight every Saturday Tonga closes down for 24 hours; the Sunday day of rest is so enshrined in Tongan culture that it's illegal to work.

1. Male humpback whales arriving in Tonga sing unique songs to attract a mate
2. On the south coast of Vava'u, Neiafu is Tonga's second-largest town, home to fewer than 4000 people
3. The king and queen of Tonga make a procession to the royal palace for the coronation in 2015
4. A traditonal Tongan house in Neiafu

1. Peak egg-laying activity occurs from March to June for leatherback turtles on Tobago

2. Masqueraders at the Port of Spain Carnival

3. On the sheltered north coast of Tobago, Englishman's Bay is one of the island's best beaches

(T) CAPITAL PORT OF SPAIN // POPULATION 1.2 MILLION // AREA 5128 SQ KM // OFFICIAL LANGUAGE ENGLISH

Trinidad & Tobago

Like a perfect callaloo (thick green stew), Trinidad and Tobago are a rich hotpot of flavours, cultures and influences. You can hear it in the music, which combines everything Caribbean (calypso, soca, steel pan, parang, reggae and much more) and see it in the locals, a mélange of African, Spanish, Asian and Amerindian peoples. Half of the capital, Port of Spain, is falling apart while the rest buzzes with an energy that can overwhelm; often you can't separate the two. Oil and gas production are Trinidad's economic engine, so tourism can be an afterthought – good news if you hate well-trodden paths. Tobago is more visitor-oriented, but the beaches still throng with as many locals as tourists. Trinidad and Tobago are always authentic.

Best time to visit
It is slightly wetter May to September; the real joy is at Carnival in February

Top things to see
- Trinidad's remote northeast coast
- Queen's Park Savannah, where all of Trinidad comes to sip coconut water and people-watch
- Brasso Seco, the rainforest village
- Trinidad's Asa Wright Nature Centre, where a full list of bird species tops 430

Top things to do
- Lose yourself at Port of Spain's world-class annual Carnival, the Caribbean's biggest and best
- Surf Mt Irvine Bay on Tobago and wheedle out tips on secret spots from locals
- Snorkel the coral gardens of Speyside in Tobago
- Dip into waterfalls and hike through the rainforest in Trinidad's coastal Northern Range

Getting under the skin
Read: *A House for Mr Biswas* by VS Naipaul – praised as one of the greatest novelists of his time, though often controversial – this is widely regarded as Naipaul's best work
Listen: to Lord Kitchener's classic calypsos or rapid-fire soca from Machel Montano
Watch: *Bazodee,* a charming romantic story about the music and culture of Trinidad
Eat: callaloo (a thick green stew made with dasheen leaves, okra, pumpkin and seasonings)
Drink: *sorrel* (made from a type of hibiscus, mixed with cinnamon and other spices)

In a word
You limin' tonight? (Are you hanging out tonight?)

Celebrate this
Carnival in Trinidad and Tobago is the largest in the Caribbean; it's officially celebrated on the Monday and Tuesday before Ash Wednesday but festivities start just after Christmas.

Random fact
Trinidad's Grande Riviere is a top turtle-watching spot, with up to 500 leatherbacks visiting nightly.

T

T CAPITAL TUNIS // POPULATION 11.7 MILLION // AREA 163,610 SQ KM // OFFICIAL LANGUAGE ARABIC

Tunisia

As one of Africa's greatest gateways to Europe, Tunisia has long been of vital geographic importance, with the Phoenicians then the Romans and Ottoman Turks all fighting for influence. The first arrivals to this Berber-inhabited land formed Carthage, a powerful city state that grew to control much of the Mediterranean's seagoing trade. Remnants of this past empire and others, including one of the best-preserved Roman colosseums in the world, litter the landscape, as do age-old examples of troglodyte and Berber architecture. Exploring them, the clamorous medinas and the wondrous dunes of the Sahara provides ample reasons to move off the beaches, which have been the biggest tourism pull in recent decades for sun-starved Europeans.

Best time to visit
November to April

Top things to see
- El Jem, the world's third-largest colosseum and one of Africa's most impressive Roman sites
- The magnificent Roman mosaics within the Bardo Museum in Tunis
- The charming, little-visited old town of Mahdia on Tunisia's coast
- Dougga's Roman theatre, one of ancient Rome's biggest and best preserved examples
- Remains of centuries-old Berber hill villages and their *ksour* (fortified granaries) west of Tataouine

Top things to do
- Wander the low-slung ruins of Carthage, one of the ancient world's greatest cities
- See time happily melt away on the mythical island of Djerba, with its near-perfect beaches and fascinating cultural mix
- Explore the great sand dunes of the Sahara atop a camel from Douz
- Lose yourself in the labyrinth of Tunis' medina
- Imagine yourself in a *Star Wars* movie amid the architecture of Ong Jemal

Getting under the skin
Read: *This Tilting World*, Colette Fellous' love letter to her homeland following the 2015 Sousse terrorist attack
Listen: to Emel Mathlouthi's song 'Kelmti Horra' (My Word is Free), which became the anthem for Tunisia's revolution, a movement that sparked the Arab Spring
Watch: *Whispering Sands* by Tunisian filmmaker Nacer Khemir, which touches on Sufism, cultural erosion and a future of uncertainty
Eat: *harissa* (a fiery red-chilli paste); *brik* (deep-fried pastry filled with egg and other delights); seafood; couscous
Drink: Turkish coffee with orange blossom or rose water

In a word
Shukran (Thank you)

Celebrate this
Lina Ben Mhenni (1983-2020): a Tunisian blogger and civil rights activist who was prominent in the 2011 revolution; feeling the youth were pushed out of the popular movement, she continued to fight for their original objectives until her death from lupus.

Random fact
George Lucas was influenced by Tunisia when filming *Star Wars: A New Hope:* Obi-Wan Kenobi's distinctive robe was directly taken from Berber clothing, and the Sandcrawler used by the Jawas to cross Tatooine (named after the Tunisian village of Tataouine) was inspired by the strange shape of Tunis' Hotel du Lac.

1. Find traditional Berber granaries in Medenine

2. The old port of Bizerte began as a Phoenician trading post before being conquered by the armies of Julius Caesar

3. Founded by Phoenicians before being ruled by Romans, Carthage was a great civilisation of antiquity and still sparks the imagination

Best time to visit
April to October

Top things to see
- The soaring interior domes of İstanbul's Aya Sofya and the Blue Mosque
- Ephesus, the best-preserved Roman city in the eastern Mediterranean
- The colossal stone statues and toppled heads marking the summit of Nemrut Dağı (Mt Nemrut)
- Safranbolu's cobblestone lanes lined with restored Ottoman merchant mansions
- The church ruins of Ani, once capital of the Armenian Kingdom, scattered across the high plateau

Top things to do
- Hike between Cappadocia's 'fairy chimney' rock formations and Byzantine rock-cut churches then float over them in a hot-air balloon at dawn
- Sail the stretch of coast east from Fethiye in a *gület* (Turkish yacht) on a Blue Cruise
- Stroll down the gleaming white calcite terraces of Pamukkale
- Trek the Teke Peninsula, between ancient city ruins, on the Lycian Way
- Kayak over the sunken city ruins around Kekova Island, near Kaş

Getting under the skin
Read: *The Time Regulation Institute* by Ahmet Hamdi Tanpınar; and *10 Minutes 38 Seconds in this Strange World* by Elif Shafak
Listen: to the *Hayvan Gibi* or *XX* albums by Turkish psychedelic-rock band Baba Zula
Watch: *Once Upon a Time in Anatolia,* directed by Nuri Bilge Ceylan; and *Mustang,* directed by Deniz Gamze Ergüven
Eat: Gaziantep's baklava, Black Sea *kuymak* (cheesy cornmeal fondue), *İskender kebap* (lamb drenched in a tomato, yoghurt and browned butter sauce) in Bursa, *gözleme* (stuffed flatbreads) and *pide* (Turkish pizza)
Drink: *çay* (tea) in tulip-shaped glasses; *rakı* (aniseed-flavoured grape brandy); or *ayran* (whipped yoghurt drink)

In a word
Hoş geldiniz (Welcome)

Celebrate this
The Neolithic T-pillars of Göbeklitepe are considered the world's oldest place of worship and are Turkey's newest Unesco World Heritage Site.

Random fact
St Nicholas (aka Santa Claus) lived and preached in Ancient Myra (today's Demre) where you can still visit the church dedicated to him.

1. Istanbul's Blue Mosque was the grand project of Sultan Ahmet I and features 260 windows

2. Whirling dervishes follow Sufism, and the Sama ceremony creates religious ecstasy

3. A chef surveys Turkish meze dishes

4. The Roman ruins on Curetes Way lead towards the heart of Ephesus, once the fourth largest city in the Roman Empire

T   **CAPITAL** ANKARA // **POPULATION** 82 MILLION // **AREA** 783,562 SQ KM // **OFFICIAL LANGUAGE** TURKISH

Turkey

Turkey holds a full hand of historic riches. Its ruins from ancient empires, forts and castles forged by sultans and crusaders, Byzantine basilicas and Ottoman-built mosques, bazaars and caravanserais are all a reminder that this country straddles the boundary between Europe and Asia in culture as well as geography. Today, Turkey's manoeuvring to position itself as a regional leader, combined with its geopolitical location, perpetually leads it into disputes with both neighbours and global powers. Internally, discord between conservative traditionalists and liberal progressives continue to divide Turkish society. None of Turkey's troubles keep visitors away for long, as the Mediterranean's beaches, a human history stretching back across millennia and its multitude of landscapes have an enduring appeal.

1

2

T CAPITAL ASHGABAT // POPULATION 5.5 MILLION // AREA 488,100 SQ KM // OFFICIAL LANGUAGE TURKMEN

Turkmenistan

Largely isolated from the rest of the world since independence from the USSR in 1990, Turkmenistan's first two decades were dominated by eccentric late President 'Turkmenbashi' (Father of the Turkmen), who set about recreating the country in his own image, using as funds one of the world's largest reserves of natural gas. The result is a fascinating hermit fiefdom of oddball sights and intriguing historical remains. Gasp at the huge Turkmenbashi Ruhy Mosque, stare in awe at the Ministry of Fairness, and be one of the few people to have glimpsed the strangest corner of Central Asia.

Best time to visit
April to May and September to November

Top things to see
• The wacky monuments of Ashgabat, including the world's largest carpet and a building-sized copy of the *Ruhnama*, the spiritual guide authored by the former president
• The ancient Seljuq capital of Merv, once the world's largest city until pulverised by the Mongols
• Konye-Urgench, the ruined 13th-century capital of the vast Khorezmshah empire
• The views of the capital from the Turkmenbashi Cableway
• Köw-Ata Lake, an underground thermal pool on the southern fringes of the Karakum desert

Top things to do
• Search for dinosaur footprints at Kugitang Nature Reserve
• Ride an Akhal-Teke horse across the foothills of the Kopet Dag mountains
• Take a 4WD trip through the desert to the burning Darvaza Gas Craters, also known as the 'Doorway to Hell'
• Splash in the Caspian Sea in the resorts near port city Turkmenbashi before catching a ferry to Azerbaijan

Getting under the skin
Read: *Sovietistan* by Erika Fatland, whose Central Asia-wide travelogue kicks off in Turkmenistan and enjoys a journalist's eye for the quirky and surreal
Listen: to *City of Love* by Ashkabad, a lilting, romantic five-piece Turkmen ensemble
Watch: Waldemar Januszczak's undercover Turkmenbashi-era documentary *The Happy Dictator*; or the August 2019 episode of *Last Week Tonight with John Oliver* that focuses entirely on Turkmenistan
Eat: *dograma*, a traditional Turkmen meal of pieces of bread, meat and onions covered in broth
Drink: a cooling cup of *chal* (sour fermented camel's milk) as the sun rises over the Karakum desert

In a word
Siz nahili? (How are you?)

Celebrate this
Turkmenistan enjoys many national holidays, including Melon Day, Horse Racing Day, Neutrality Day, Carpet Day, Turkmenbashi Remembrance Day and Drop of Water is a Grain of Gold Day.

Random fact
Ashgabat holds Guinness world records for the 'highest concentration of white marble buildings' and the 'highest number of fountains in a public place'.

1. The burning gas crater in the Karakum desert was started by Soviet scientists who believed that it would burn out quickly: 40 years later it hasn't stopped
2. The walls of the archive of the Communist Party of Turkmenistan feature modernist concrete sculptures made by the Russian artist Ernst Neizvestny
3. Leader Gurbanguly Berdymukhamedov is a cycling enthusiast and supports World Bicycle Day
4. Honey for sale in Ashgabat's Tikinske Bazaar

T CAPITAL FUNAFUTI // POPULATION 11,342 // AREA 26 SQ KM // OFFICIAL LANGUAGES TUVALUAN & ENGLISH

Tuvalu

Tuvalu is one of the smallest, remotest and most low-lying nations on earth: its highest point is only 4.6m (15ft) above sea level. Approaching the island by plane after soaring over endless ocean, a dazzling smear of turquoise and green appears, ringed with coral and studded with coconut palms. On the ground, experience the slow pace of the unspoiled South Pacific: stroll around town, hang out at the airstrip (the island's social hub) or float in the sparkling lagoon. The most energetic activity is motorbiking up and down the country's only tarmac road on Funafuti with the local boy racers. Tuvalu's very existence is threatened by rising sea levels, an effect of climate change.

Nanumea Atoll

Nanumaga *Niutao* SOUTH PACIFIC OCEAN

Nui Atoll *Vaitupu*

Nukufetau Atoll

Funafuti Atoll
✪ Funafuti

SOUTH PACIFIC OCEAN

Nukulaelae Atoll

Niulakita

Best time to visit
May to October, the dry season, when cooling trade winds provide natural air conditioning

Top things to see
• Performance of a *fatele*, a Tuvaluan dancing song with building percussive rhythms and crescendo singing
• *Te ano*, a sport unique to Tuvalu, where two balls get hit around by co-ed teams
• Sublime sunsets while floating in the luminous waters of Funafuti Lagoon

Top things to do
• Realise your desert-island fantasies on the sparkling islets of the Funafuti Conservation Area
• Experience traditional life on a remote island with the few remaining families of Funafala Islet
• Hire a bicycle and explore the length of Fongafale
• Maroon yourself on the outer islands where the supply ship might (or might not) pick you up again in a few weeks

Getting under the skin
Read: *Where the Hell is Tuvalu?*, a comic tale by Philip Ells about being in Tuvalu as 'the people's lawyer'; *Time and Tide: The Islands of Tuvalu*, a photographic essay by Peter Bennetts and Tony Wheeler
Listen: to Te Vaka, a mesmerising group of dancers and musicians fusing traditional Pacific and contemporary sounds
Watch: eye-opening documentary *ThuleTuvalu*, which looks at how melting ice in Greenland threatens to drown tiny Tuvalu
Eat: a streetside snack of chicken curry or fish roti
Drink: coconut water straight from the nut

In a word
Fifilemu (To be very peaceful, quiet)

Celebrate this
Tuvalu may be small...but it sure is savvy. Running low on national funds in the late 1990s, Tuvalu leased its top-level domain suffix – .tv – for US$50 million; it's now set to renew the deal for even more money.

Random fact
'Tuvalu' means 'eight standing together'; it was named for the nation's eight continuously inhabited islands.

1. Tuvalu consists of 124 islands, islets and atolls

2. This narrow strip of land in Funafuti highlights how vulnerable Tuvalu is to rising oceans

3. A beachside cafe serves evening drinks

4. During high tides, low-lying areas often flood; sea water sometimes percolates up through the coral

Best time to visit
January to February or June to September

Top things to see
- Herds of elephants, pods of hippos, prides of lions and towers of giraffes in Queen Elizabeth National Park
- A narrow gorge trying to strangle the mighty Nile at Murchison Falls, resulting in the world's most powerful waterfall
- Terraced hills flowing down to Lake Bunyonyi
- The white-sand beaches of the Ssese Islands
- Candlelit vegetable stalls punctuating the darkness on Kampala's backstreets

Top things to do
- Wade into Bwindi's Impenetrable Forest National Park to come face-to-face with mountain gorillas
- Stir the Nile with your paddle before it shakes you (and your raft) to the core at Kalagala Falls
- Trek through the cold to reach mystical highs in the glaciated 'Mountains of the Moon' (Rwenzoris)
- Listen to the chorus of chimp pant-hoots while following in their actual footsteps in Kibale Forest National Park
- Track lions and cheetahs in Kidepo Valley National Park, one of Africa's wildest protected areas

Getting under the skin
Read: *Kintu* by Jennifer Nansubuga Makumbi – Uganda's history reimagined; *Waiting* by Goretti Kyomuhendo, which speaks of rural rituals and the fallout of Idi Amin

Listen: to the rumba sounds of Afrigo Band, Uganda's longest-running group; for new music check out the nation's top DJ, DJ Erycom

Watch: *The Last King of Scotland*, Kevin Macdonald's adaptation of Giles Foden's novel about the emotional whirlwind surrounding Idi Amin's physician; *Queen of Katwe*, the story of a chess phenom rising from Kampala

Eat: *matoke* (cooked plantains) and groundnut sauce; *rolex*, a chapatti wrapped around an omelette

Drink: Chairman's ESB beer; or less tame is *waragi* (millet-based alcohol)

In a word
Habari? (What news? In Swahili)

Celebrate this
Christopher Ategeka: an engineer and entrepreneur who founded Rides for Lives, a non-profit healthcare provider that serves rural Ugandan communities (now Health Access Corps, it works across sub-Saharan Africa); he then founded Hourglass Ventures to support underprivileged African entrepreneurs.

Random fact
Winston Churchill was one of Uganda's earliest tourists – his 1907 visit had a lasting impact, with his 'Pearl of Africa' description becoming one of the nation's monikers.

1. A silverback mountain gorilla, one of about 300 living in Bwindi Impenetrable Forest National Park

2. Downtown Kampala, a city notorious for its traffic jams

3. Murchison Falls are formed as the Victoria Nile River squeezes through a rocky gorge

4. Looking over Bwindi Impenetrable Forest National Park towards the Virunga Mountains of Rwanda

U **CAPITAL** KAMPALA // **POPULATION** 43.3 MILLION // **AREA** 241,038 SQ KM // **OFFICIAL LANGUAGE** ENGLISH

Uganda

An amalgamation of four African kingdoms, and home to a multitude of peoples speaking some 43 languages, Uganda takes its name from its largest kingdom, the Buganda. As well as cultural diversity, the nation also reverberates with nature: dense forest canopies shield the playground of hundreds of bird and mammal species, including half the planet's mountain gorillas; savannahs, nestled in some of Africa's most stunning settings, nurture classic safari wildlife; and the Rwenzoris, Africa's highest mountain range, are clad in endemic plant species and topped with equatorial glaciers. Although controversial anti-gay legislation has cast a shadow on the entire nation, most Ugandans are open and warm-spirited.

DANITA DELIMONT STOCK | AWL IMAGES // JON ARNOLD | AWL IMAGES // HANS NELEMAN | GETTY IMAGES // JON ARNOLD | AWL IMAGES

U CAPITAL KYIV // POPULATION 43.9 MILLION // AREA 603,550 SQ KM // OFFICIAL LANGUAGE UKRAINIAN

Ukraine

Ukraine is the birthplace of Eastern Slavic civilisation, with a capital at least 1500 years old, and home to seminomadic Hutsuls who still inhabit the Carpathian Mountains. Experiencing this enormous country can mean anything from Orthodox churches to mountain passes, packed beaches or pulsating nightlife. Having weathered a political revolution, the capital, Kyiv, is bursting with newfound creativity. Lviv is the rising tourist hot spot, with its period architecture and traditional coffee houses, while the Black Sea port of Odesa has long been a cultural melting pot. In 2014 Russia's annexation of Crimea, the peninsula jutting from Ukraine's southeast, started another turbulent chapter in a region already marked by political instability. Certainly, Ukraine's story is far from concluded.

Best time to visit
May to October

Top things to see
- The caverns with mummified remains of monks at Kyevo-Pecherska Lavra
- Folk art, windmills and traditional village huts at the open-air Museum of Folk Architecture in Pyrohiv
- The fortress of Kamyanets-Podilsky in an impressive setting above the Smotrych River
- Carpathian National Park, Ukraine's largest wilderness preserve and home to lynx, bison, brown bears and wolves

Top things to do
- Hop aboard a boat tour in the Danube Delta Biosphere Reserve, a haven for great white pelicans, red-breasted geese and other feathered friends
- Stroll the seaside promenade in Odesa, taking in the cinematic icon of the Potemkin Steps, and catch a show at the Opera and Ballet Theatre
- Lose yourself in the atmospheric back lanes of Lviv, with buildings dating back to the 14th century and intimate coffee houses
- Delve into Kyiv: marvel at 1000-year-old St Sophia Cathedral, shop along cobblestone Andriyivsky Uzviz and join in the hipster bar scene

Getting under the skin
Read: *Death and the Penguin* by Andrey Kurkov, an absurdist and socially incisive novel about a man and a penguin living in modern-day Kyiv
Listen: to the fast-paced dance melodies of traditional folk-music ensemble Suzirya
Watch: Oksana Bayrak's *Avrora*, the evocative story of a 12-year-old orphan and aspiring dancer who witnesses the Chornobyl disaster
Eat: tasty *varenyky* (boiled dumplings served with cheese or meat); borsch (beetroot soup); and *salo* (salted or smoked pork fat)
Drink: vodka; *kvas* (a sweet beer made from fermented bread)

In a word
Vitayu (Hello)

Celebrate this
The Carpathian-dwelling Hutsul ethnic group is a mainstay of Ukrainian national identity. Two museums dedicated to Hutsul folk art and way of life, in Kolomyya and Kosiv, have well-curated exhibitions.

Random fact
By all accounts you are exposed to twice the amount of radiation on a trans-Atlantic flight than you are during an entire day at Chornobyl.

1. Pripyat amusement park was due to open the week after the Chernobyl disaster occured

2. St Sophia's cathedral in Kyiv contains frescoes dating back to 1017

3. Ballet is central to Ukrainian culture and its schools are some of the best in the world

4. The Motherland Monument at the National Museum of the History of Ukraine in WWII was sculpted by Vasyl Borodai

1. The traditional way to cross Dubai Creek in the old part of the city is by *abra* (water taxi)

2. The road to the top of Abu Dhabi's highest mountain, Jebel Hafeet, is a favourite with drivers and cyclists

3. The five Etihad Towers in Abu Dhabi appeared in a *Fast and Furious* movie in 2015

U | **CAPITAL** ABU DHABI // **POPULATION** 9.7 MILLION // **AREA** 83,600 SQ KM // **OFFICIAL LANGUAGE** ARABIC

United Arab Emirates

Few countries have undergone such a super-charged transformation as the seven emirates of the UAE. Until oil was discovered in the late-1950s, this was a sparsely populated Gulf backwater with Dubai and Abu Dhabi simple fishing settlements. In less than a century, extraordinary oil wealth has allowed the Emirati ruling families to remodel the UAE into a multicultural (immigrants make up around 80% of the population) business hub with burgeoning regional leader ambitions while still strictly rooting UAE society in traditional Islamic values. Though mostly known for city-slicker Dubai and Abu Dhabi, off-the-grid adventures beckon in the Hajjar Mountains and Empty Quarter desert dunes, while cultural heritage from before the oil-boom is preserved in Al Ain and Sharjah.

Best time to visit
October to April

Top things to see
- Dubai's skyline from the Burj Khalifa
- Abu Dhabi's Sheikh Zayed Grand Mosque and Louvre Abu Dhabi museum
- Bronze Age tombs and caves used in the Palaeolithic era, surrounded by desert, at Mleiha Archaeological Site
- Sharjah's Art Museum and restored heritage district
- Jebel Hafeet's summit views, Al Jahili Fort and the tombs of Hili Archaeological Park in Al Ain

Top things to do
- 4WD from Liwa Oasis into the sand dunes of the Rub' Al-Khali (Empty Quarter) to camp overnight
- Brave the world's longest zip line at Jebel Jais
- Explore Hatta's mountain-bike trails then kayak on Hatta Dam
- Get up close and personal with raptors on a tour of Abu Dhabi Falcon Hospital
- Kayak in Mangrove National Park in Abu Dhabi

Getting under the skin
Read: *The Sand Fish* by Maha Gargash; and *Temporary People* by Deepak Unnikrishnan
Listen: to the soul-electronica infused self-titled EP from Dubai singer Hamdan Al-Abri
Watch: *Shabab Sheyab (On Borrowed Time)*, directed by Yasir Al Yasiri
Eat: *khuzi* (stuffed whole roast lamb); and *luqaimat* (deep-fried pastry balls drenched in honey)
Drink: cardamom-infused *qahwa* (Arabic coffee)

In a word
Al-Hamdu lillah (Thanks be to God)

Celebrate this
In 2020 the UAE Space Agency successfully launched unmanned probe *Al-Amal* (Hope) on its two-year mission to Mars to study the planet's atmosphere.

Random fact
Excavations at Mleiha Archaeological Site revealed the earliest signs of human habitation outside of Africa.

U CAPITAL WASHINGTON, DC // POPULATION 332.6 MILLION // AREA 9.8 MILLION SQ KM // OFFICIAL LANGUAGE ENGLISH

United States of America

The USA is a country in perpetual motion. States were formed and united by people who emigrated here, encountering Native Americans along the way. It's given us wagon trains and road trips, red-eye flights and moon landings. And given the natural beauty and iconic cities, it's no wonder locals and visitors continue to travel to explore this huge, diverse country. New York and Los Angeles are global cultural hubs, with San Francisco, Chicago and Boston close on their heels. Yosemite, Yellowstone and Grand Canyon national parks show off nature at its most sublime. And the food is as eclectic as the population, an A–Z of global cuisine alongside world-conquering chains. Buckle up, it's going to be a fantastic ride.

Best time to visit
Year-round

Top things to see
- New York City, which locals already assume you know is the centre of the universe
- California, which combines boundless beauty with the world's fifth-largest economy
- The jaw-dropping vistas of the Grand Canyon stretching into infinity
- New Orleans, where the US meets France, Africa and the Caribbean
- The natural fireworks of autumn leaves in New England

Top things to do
- Hit the museums and splendid government buildings of Washington, DC
- Escape the hordes on a back trail in Yellowstone National Park
- Joyfully join the smiling masses at Disneyland
- Road trip through the South, land of blues and barbecue
- Bike across San Francisco's Golden Gate Bridge

Getting under the skin
Read: Jack Kerouac's *On the Road*; Harper Lee's *To Kill a Mockingbird*; F Scott Fitzgerald's *The Great Gatsby*
Listen: to Bob Dylan's poetic *Highway 61 Revisited*; country great Johnny Cash's *Live at Folsom Prison*; *King of the Delta Blues Singers* by blues legend Robert Johnson
Watch: American classics: *Citizen Kane, Some Like It Hot, The Godfather, Blade Runner, Forrest Gump, 12 Years a Slave, Moonlight*
Eat: New York pizza; Chicago hot dogs; Southern barbecue; Maine lobster; Tex-Mex burritos; apple pie
Drink: craft beers; Californian reds; regional soft drinks like Moxie, Cheerwine and Dr Brown's

In a word
Hey! Howdy! Hi!

Celebrate this
The USA's treasured national parks include some of the oldest in the world (Yosemite and Yellowstone), along with North America's highest peak (Alaska's Denali) and lowest and hottest points (California's Death Valley).

Random fact
Mammoth Cave National Park in Kentucky contains the world's largest known cave system.

1. The 16-mile coastline of Hawaii's Na Pali Coast Wilderness State Park is Kaua'i's most magnificent natural sight

2. Gotham City, the Big Apple, from the Top of the Rock Observation Deck on the Rockefeller Center

3. Near Page in Arizona, the Colorado River cut through sandstone over aeons to create Horseshoe Bend

4. A jazz choir sings in St Augustine Catholic Church in New Orlean's Tremé district

U

U CAPITAL MONTEVIDEO // POPULATION 3.4 MILLION // AREA 176,215 SQ KM // OFFICIAL LANGUAGE SPANISH

Uruguay

By almost every quality-of-life index – high literacy, low corruption, freedom of press and a host of other civil liberties – ultra-liberal Uruguay is a Latin American leader. This verdant nation shares with its bigger neighbours Brazil and Argentina a passion for football, a proclivity for juicy home-grown steak and a thriving gaucho culture of horsemen, cattle ranches and big open skies. Peaceful and prosperous, Uruguay and its rolling farmland rear up in a charming coastline of colonial towns, fishing villages, surf beaches, nature reserves and – in Punta del Este – probably South America's liveliest beach party resort.

Best time to visit
December to March (summer)

Top things to see
- Montevideo, Uruguay's culturally rich capital with 19th-century neoclassical buildings and a photogenic Old Town
- The picturesque cobblestone streets of Colonia del Sacramento, beautifully set above the Río de la Plata
- Cabo Polonio, a fishing village that attracts a staggering amount of wildlife – sea lions, seals and penguins, with whales spotted offshore
- The laid-back, backpacker-friendly former fishing village of Punta del Diablo

Top things to do
- Discover Uruguay's cowboy culture at the Fiesta de la Patria Gaucha (March) in Tacuarembó, featuring rodeos, parades and folk music
- Join the international party crowd at the dance clubs in Punta del Este
- Tour the gloomy ruins of the infamous Fray Bentos meat-processing factory, a Unesco World Heritage Site
- Enjoy a soak in the thermal baths of Termas de Daymán
- Feast on the local beef at a *parrilla* (steakhouse) inside Montevideo's Mercado del Puerto

Getting under the skin
Read: any of Juan Carlos Onetti's short stories or novels set in the fictional town of Santa Maria
Listen: to singer-songwriter and Academy Award–winner Jorge Drexler; or ska-touched rockers No Te Va Gustar
Watch: *A Twelve-Year Night*, Álvaro Brechner's stark, well-crafted tale of the imprisonment of the influential leaders of the Tupamaros guerilla group
Eat: a *chivito*, a huge steak sandwich stuffed with cheese, tomatoes, bacon and more
Drink: maté, the smooth tea made from the leaves of yerba maté, sometimes served in a hollow gourd with *bomba* (a metal straw)

In a word
¿Me estás jodiendo? (Are you kidding me?)

Celebrate this
Uruguay consistently tops the continental rankings for prosperity, freedom and democracy. Almost all its energy is renewably sourced, and it's been rated as the fifth most welcoming place for gay travellers in the world.

Random fact
In 2013 Uruguay became the world's first country to legalise the cultivation and sale of marijuana.

1. Punta del Este is a sybaritic resort-city on the coast, where the wealthy pay to play
2. Gauchos ride untamed horses at a criollo festival in Canelones
3. The towers of the Basilica of the Holy Sacrament in Colonia del Sacramento
4. A vintage Studebaker awaits its next ride in the streets of Colonia del Sacramento's old quarter

1. The three medressas that form Registan Square in Samarkand are the centrepiece of the city, built between 1400 and 1650

2. A silk worker in Fergana region continues the tradition that put Samarkand on the Silk Road map

3. A photography exhibition inside the Tilla-Kari medressa illustrates how derelict the buildings were at the start of the 20th century before the Soviets started to restore them

U CAPITAL TASHKENT // POPULATION 30.6 MILLION // AREA 447,400 SQ KM // OFFICIAL LANGUAGE UZBEK

Uzbekistan

Any Silk Road romantic who has daydreamed of following the Golden Road to Samarkand or caravan trails to Bukhara will already have their sights firmly set on Uzbekistan. As the cultural and historic lynchpin of Central Asia, the country's Islamic architecture of floating turquoise domes and towering minarets ranks as the region's greatest draw, while in the foreground old men with white beards and stripy cloaks haggle over melons in the bazaar or savour a pot of green tea beside a crackling kebab stand. Improvements in red tape, visa-free travel and streamlined currency exchange have revolutionised travel here in recent years, making now the perfect time to explore the heart of Central Asia.

Best time to visit
April to June, September to October

Top things to see
- Samarkand's Registan Square, one of the world's great architectural ensembles
- Shah-i-Zinda, Samarkand, a necropolis that marks the highpoint of Timurid tilework
- The old slave-raiding walled city of Khiva, frozen in time architecturally and surrounded by desert
- The Savitsky Museum, a collection of Russian avant-garde art in the remote city of Nukus
- Bukhara's Kalon Minaret, whose beauty and scale impressed even Genghis Khan

Top things to do
- Explore the millennia-old citadels of Khorezm
- Haggle for *suzani* embroidery and tie-dyed *ikat* silks in backstreet bazaars
- Ride the Soviet-era Tashkent metro with its lavishly decorated stations
- Get lost in the mosques, mausoleums and madrasas of Bukhara's atmospheric backstreets

Getting under the skin
Read: Peter Hopkirk's *The Great Game* to get familiar with the region's imperial shenanigans
Listen: to *Yol Boisin*, by Sevara Nazarakhan
Watch: *The Desert of Forbidden Art*, tracing Igor Savitsky's quest to create an oasis of Soviet art in end-of-the-world Karakalpakstan
Eat: *plov*, a heavy dish of rice, lamb and carrots
Drink: a *piala* (bowl) of *kok choy* (green tea) at a traditional *chaikhana* (teahouse)

In a word
Yol boisin (May your travels be free of obstacles)

Celebrate this
Political and economic reforms following the death of dictator Islam Karimov in 2016 have transformed Uzbekistan society, turning it from political pariah to a tentative beacon of hope in the region.

Random fact
Uzbekistan is one of two countries to be double landlocked (two countries from the sea).

Vanuatu

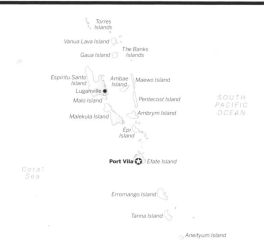

An ancient living culture, accessible volcanoes, world-class diving and some of the best cuisine in the Pacific make Vanuatu an extraordinary place to visit. Despite a history of indentured labour (known as blackbirding) and conjoint colonising by the French and British, islanders welcome visitors with authentic warmth. Smiles are as easy to come by as *laplap* cooked in a pit oven, crusty French bread or fresh fish flavoured with herbs and coconut. Port Vila, the compact capital and tourist centre, is old-school tropical-cool with glorious views of the harbour. From here set off to the outer islands for perfect beaches and crystal-clear waters – as well as black magic, fiery eruptions of lava and authentic Melanesian *kastom* (custom).

Best time to visit
April to October (winter), when temperatures aren't too stifling

Top things to see
- The explosive volcanic fireworks of Mt Yasur
- The land divers of Pentecost, who leap from man-made wooden towers
- The Dog's Head cannibal site, complete with dismemberment tables and fire pits
- Rom dances, sorcery and cultural demonstrations at one of north Ambrym's festivals
- The chambers, tunnels and underground lake of Valeva Cave from a kayak

Top things to do
- Camp by the active caldera of Mt Marum surrounded by lava beds, jungle and cane forests
- Swim in an underwater world of sunken luxury liners, caves and coral gardens off Luganville
- Keep your eyes peeled for dugongs while snorkelling in Epi's Lamen Bay
- Relax 'resort style' at the delightful Oyster Island on Santo
- Splash about in the aquamarine freshwater pool at the base of the 35m Mele Cascades

Getting under the skin
Read: *Getting Stoned with Savages*, a witty account of J Marten Troost's misadventures in Vanuatu and Fiji; *Things Bright and Beautiful* by Anbara Salam, a suspenseful dark comedy set in postwar New Hebrides (as Vanuatu was known)
Listen: to *Best Of* by the ni-Vanuatu singer Vanessa Quai, a 'living cultural icon'
Watch: the 2015 movie *Tanna*, shot in Vanuatu and starring the indigenous Yakel tribe
Eat: *laplap*, taro paste cooked with coconut and meat in an earthen oven
Drink: *aelan bia* (island beer), aka kava, a becalming non-alcoholic but narcotic brew

In a word
Tank yu tumas (Thank you very much)

Celebrate this
Vanuatu is ranked fourth on the Happy Planet Index, which measures sustainable well-being around the world. Vanuatu scored highly for its peacefulness, the importance placed on community engagement, and its low ecological footprint.

Random fact
Yaohnanen villagers on Vanuatu's Tanna Island worship Prince Phillip, Duke of Edinburgh, believing he is a divine being and the son of a mountain god.

1. A lava lake simmers inside Mt Marum's active crater on Ambrym island
2. The Rom dance on Ambryn, Vanuatu's spiritual core, sees dancers wear a painted *rom* mask and a *rablar* coat made from banana leaves
3. Minuscule Mystery island lies just off Aneityum, Vanuatu's southernmost island
4. There are more than 20 active or extinct volcanoes that are part of the Vanuatu archipelago, many uninhabited

(V) POPULATION 839 // AREA 0.44 SQ KM // OFFICIAL LANGUAGES LATIN & ITALIAN

Vatican City

It is quite extraordinary to find a fully functioning city-state in the middle of a modern European capital. The chaos and attitude of Rome ends at the walls of the Vatican, where medieval traditions and arcane rituals take the place of coffee-shop culture and Vespas. This – the world's smallest nation – is the seat of the Catholic Church, ruled with absolute authority by the Pope, from a palace with over a thousand rooms, protected by a squadron of single Swiss men in flamboyant red, yellow and blue uniforms. You don't have to be Catholic to appreciate the Vatican; the city-state is also one of the world's great repositories of architecture, culture and art.

Best time to visit
April and June (the low season); Wednesdays when the Pope meets his flock

Top things to see
- Michelangelo's Sistine Chapel – his ceiling frescoes and *The Last Judgement* are awe-inspiring
- The Pope, who turns up consistently at 10.30am on Wednesdays (get there early to get a seat!)
- The Swiss Guard – all male, Swiss and single – and their marvellous, Renaissance-styled skirt pants
- One of the world's greatest collections of sacred art inside the Vatican Museums

Top things to do
- Hike the 320 steps up St Peter's Basilica, the world's largest dome – the panorama is dizzying and dazzling
- Zigzag around Doric columns and absorb the extraordinary air of St Peter's Square
- Kiss or rub for luck the right foot of bronze St Peter inside the basilica
- Take a tour of the 'City of the Dead' and see St Peter's tomb beneath the basilica
- Experience a calmer side to the Vatican in the Vatican Gardens, a lavish sprawl of lawns, grottoes, fountains and topiary

Getting under the skin
Read: *A Season for the Dead* by David Hewson for a fast-paced thriller about a serial-killer cardinal
Listen: to papal speeches and news from the city on Radio Vatican (live-streaming or podcasts in English) at www.radiovaticana.org
Watch: Tom Hanks flit around the Vatican City in Ron Howard's film adaptation of Dan Brown's *Da Vinci Code* sequel, *Angels and Demons*
Eat: Roman pasta such as creamy carbonara (egg yolk, parmesan and bacon); and fiery *alla matriciana* (tomato, bacon and chilli)
Drink: wines such as Frascati and Torre Ercolana (the Vatican consumes more wine per person than any other country)

In a word
Amen

Celebrate this
In 2019, Pope Francis unveiled a monument in St Peter's Square dedicated to the world's refugees and featuring 140 migrants, including Mary and Joseph, the parents of Jesus.

Random fact
The Popes' previous pad was in Avignon, southern France; the clergy moved to the Vatican City in 1377.

1. This is the modern tribute to the famous Bramante spiral staircase in the Vatican Museums, dating from 1932

2. St Peter's is Italy's largest and most spectacular basilica, a symbol of the Catholic Church's wealth

3. The Pontifical Swiss Guard has both ceremonial and protective roles

1. The idyllic Caribbean archipelago Los Roques, with 300 islands, is a dependency of Venezuela

2. Around two million people live in the Venezuelan capital, Caracas, surrounded by coastal mountains

3. Angel Falls plummets 979m as the Río Kerepacupai Merú flows off Auyán-tepui mountain

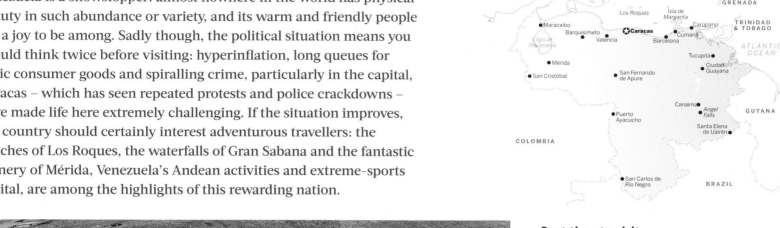

Ⓥ **CAPITAL** CARACAS // **POPULATION** 28.6 MILLION // **AREA** 912,050 SQ KM // **OFFICIAL LANGUAGE** SPANISH

Venezuela

Venezuela is a showstopper: almost nowhere in the world has physical beauty in such abundance or variety, and its warm and friendly people are a joy to be among. Sadly though, the political situation means you should think twice before visiting: hyperinflation, long queues for basic consumer goods and spiralling crime, particularly in the capital, Caracas – which has seen repeated protests and police crackdowns – have made life here extremely challenging. If the situation improves, the country should certainly interest adventurous travellers: the beaches of Los Roques, the waterfalls of Gran Sabana and the fantastic scenery of Mérida, Venezuela's Andean activities and extreme-sports capital, are among the highlights of this rewarding nation.

Best time to visit
December to April (the dry season)

Top things to see
- Angel Falls, Earth's highest waterfall, dropping 300 storeys in Parque Nacional Canaima
- Los Roques, tiny islands with a friendly vibe
- Los Llanos, a savannah with anteaters, capybaras, anacondas, caimans and astounding birdlife
- Puerto Colombia in the Parque Nacional Henri Pittier, with its magnificent selection of beaches
- Gran Sabana's table mountains, roaring rivers, gushing waterfalls and wilderness

Top things to do
- Go paragliding, canyoning, mountain climbing and hiking in the highland city of Mérida
- Dive the waters of Los Roques archipelago
- Hike to the top of Roraima, a massive table mountain with a wild landscape
- Explore the labyrinthine channels of the wildlife-packed Orinoco Delta
- Discover the magnificent and largely undeveloped beaches of the Península de Paria

Getting under the skin
Read: *Doña Bárbara* by Romulo Gallegos, a tale of power and progress on the grasslands
Listen: to the disco-funk of Los Amigos Invisibles
Watch: *The Revolution Will Not Be Televised*, a documentary about the failed coup to oust former president Hugo Chávez
Eat: *pabellón criollo* (shredded beef, rice, black beans, cheese and fried plantain)
Drink: *guarapita* (sugar-cane spirit and fresh juices)

In a word
Como esta? (How are you?)

Celebrate this
The Orinoco River snakes for 2250km, from the Colombian highlands to the vast Orinoco Delta.

Random fact
Sir Arthur Conan Doyle's book *The Lost World* was inspired by the table mountain of Roraima, where unique species were discovered in the 1880s.

 CAPITAL HANOI // POPULATION 98.7 MILLION // AREA 331,210 SQ KM // OFFICIAL LANGUAGE VIETNAMESE

Vietnam

The story of Vietnam is a tale of two cities: historic Hanoi and have-a-go Ho Chi Minh City. The slender arc of land strung out between these former wartime rivals has been the setting of some of Asia's greatest dramas, and some of its greatest tragedies, but it's a mistake to view Vietnam solely through the prism of war. Rather than dwelling in the past, the Vietnamese are moving forcefully towards the future, riding on a wave of technological innovation. For travellers, the geography encourages a linear trajectory, flying into Hanoi or Ho Chi Minh, then meandering past paddy fields, white-sand beaches, jungles, hill-tribe villages and sculpted limestone islands. En route, you may get whiplash from the pace of change in Asia's latest hi-tech hub.

Best time to visit
March and April, September to November

Top things to see
- Hanoi's temple-topped Hoan Kiem Lake, lined with morning exercisers at dawn
- The unmistakably communist lines of the Ho Chi Minh Mausoleum in Hanoi
- Dragon-shaped mountains and jade-green waters in Halong Bay
- The citadel and royal tombs strung along the Perfume River in Hué
- Terraced rice fields, forested mountains and traditional tribal homes around Sapa

Top things to do
- Enjoy *pho* (noodle soup) at a sidewalk eatery anywhere in Vietnam
- Take slow boats though the backwaters of the Mekong Delta
- Learn to make your own *goi cuon* (soft summer rolls) on a cooking course in Hoi An
- Explore the countryside by motorcycle, and discover Vietnam's gentler, calmer side
- Experience the great outdoors – on foot, by mountain-bike or on your hands and knees in the tunnels excavated by the Viet Cong

Getting under the skin
Read: *Dumb Luck* by Vũ Trọng Phụng, a satire of colonial-era social mores; or Bảo Ninh's *The Sorrow of War*, unveiling the Vietnamese view of the conflict
Listen: to the crowd-pleasing pop of Sơn Tùng M-TP, or the haunting sound of *ca trù*, musical storytelling form from northern Vietnam
Watch: Đặng Nhật Minh's *When the Tenth Month Comes*, exploring the war from the Vietnamese perspective; or Nhất Trung's *Cua Lai Vo Bau*, a rom-com that became Vietnam's biggest ever grossing film in 2019
Eat: *pho* soup and *banh cuon* (steamed rice rolls with minced pork)
Drink: *bia hoi* (draught beer), or, for the brave, rice wine containing pickled snakes and scorpions

In a word
Troi oi! (Oh my!)

Celebrate this
Model H'Hen Nie, from the Ede tribal group, has used victory in pageants as a platform to campaign against child-marriage and bias in education.

Random fact
Superglue was reputedly invented as an emergency wound dressing in the Vietnam War.

1. The Reunification Express runs for 1000 miles beside the South China Sea between Hanoi and Ho Chi Minh City
2. Rice terraces around the village of Cat Cat in the northern highlands provide food for local tribal people
3. Two wheels rules the roads of Hanoi, in Vietnam's north

W **CAPITAL** CARDIFF // **POPULATION** 3.2 MILLION // **AREA** 20,764 SQ KM // **OFFICIAL LANGUAGE** WELSH & ENGLISH

Wales

Wales at its best is hiking in splendid isolation across magnificent green hills and purple-heather moors, valleys ringing with the song of male voice choirs and poetry of 6th-century bards, local pubs beckoning with a pint of bitter. This raw, underrated land of dramatic landscapes, mighty stone castles and atmospheric morning mists is enriched by myths, literature and the fiercely patriotic, rugby-loving, song-mad Cymry (Welsh) themselves. Rejuvenated modern cities and a grand industrial heritage lie alongside all that natural beauty, while an ever-increasing range of adventure sports and some fine festivals keep this small, damp, glorious nation buzzing.

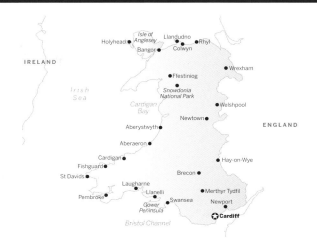

Best time to visit
May to September

Top things to see
• The towering, elegant, 200-year-old aqueduct at Pontcysyllte
• Snowdon, Wales' highest peak at 1085m
• A rugby match at the Principality Stadium in Cardiff
• Conwy, Caernarfon, Harlech and Beaumaris castles
• The Brecon Beacons National Park – rugged hills, moors and fantastic pubs

Top things to do
• Hike, mountain bike, camp, trampoline and surf artificial waves in Snowdonia National Park
• Shop for china in Portmeirion, a whimsical vision of Italian classicism
• Listen to male voice choirs at Llangollen's International Eisteddfod and folk and rock at Crickhowell's Green Man
• Follow the Pembrokeshire Coast Path through quaint fishing villages and around secluded coves
• Frolic across sandy beaches and limestone cliffs on the Gower Peninsula

Getting under the skin
Read: *Random Deaths and Custard* by Catrin Dafydd, one of Wales' best contemporary writers; *The Mabinogion*, medieval tales from Wales
Listen: to operatic arias by Welsh tenor Aled Wyn Davies; male voice choirs from 'the land of song'
Watch: *Solomon & Gaenor*, a turn-of-the-20th-century tale of forbidden love set against South Wales' coalfields
Eat: *bara brith* (tea-soaked fruit loaf); Welsh rarebit (beer-soaked cheese on toast); laverbread (misleadingly named seaweed dish); vegetarian-friendly Glamorgan sausages
Drink: a pint of Cardiff-brewed Brains or ale from a local microbrewery

In a word
Bore da (Hello, good morning)

Celebrate this
Owain Glyndŵr, the last Welsh-born Prince of Wales, has been a national hero since his legendary, though unsuccessful, battles to kick out the occupying English in the early 15th century.

Random fact
Tiny St Davids is the UK's smallest city, home to just 1600 people – plus the tomb of the eponymous saint.

1. It's only the fifteenth-highest Welsh mountain but you'll need scrambling skills to summit Tryfan in Snowdonia

2. Engineer Thomas Telford (1757–1834) built the Pontcysyllte Aqueduct in 1805 to carry the canal over the River Dee

3. Look out over the tip of the Pembrokeshire coast in the southwest of Wales

Y CAPITAL SANA'A // POPULATION 29.8 MILLION // AREA 527,968 SQ KM // OFFICIAL LANGUAGE ARABIC

Yemen

With none of the oil wealth of its neighbours, parts of Yemen always seemed like a time capsule to old Arabia, from Sana'a's historic rammed-earth houses to the unique ecology and landscapes of Socotra Island. Yemen's history reads like the retelling of legends – Noah is said to have launched his ark from here, the Queen of Sheba once ruled the land and there was dazzling wealth from the frankincense trade. Today, Yemen's fortunes are about as distant from those stories as it's possible to be. With civil war, and famine conditions in the worst-hit regions, Yemen has been called the worst humanitarian crisis in the world.

Best time
October to March

Places of interest
- Sana'a, the 2500-year-old Unesco World Heritage–listed city with 14,000 ancient buildings
- The mudbrick high-rise architecture of Shibam in Wadi Hadramawt
- Socotra Island, for its distinct ecology, landscapes and culture
- The historic fortified village of Thilla
- The fertile highland scenery surrounding the towns of Ibb and Jiblah

Local activities
- Treks heading inland to Homhil and Firmhin Forest on Socotra Island offer landscapes of the endemic dragon's blood trees
- The drive up to Shaharah, and the ancient mountain village itself with its 17th-century suspended bridge, afford rolling mountain vistas
- Sana'a's Souq Al-Milh is the old city's hub of local produce and spice shopping
- The *bara* is Yemen's famous traditional dagger dance
- Chewing qat (the leaves of a mildly narcotic plant) is a national pastime

Getting under the skin
Read: *Hurma* by Ali Al-Muqri; and *Yemen: Travels in Dictionary Land* by Tim Mackintosh-Smith
Listen: to *Habibi Ta'al* by Ahmed Fathi, a renowned Yemeni oud player and singer
Watch: *10 Days Before the Wedding*, directed by Amr Gamal; and documentaries *Karama Has No Walls* and *The Mulberry House,* both directed by Sara Ishaq
Eat: *saltah* (meat stew flavoured with tomatoes, chillies and fenugreek)
Drink: tea scented with cardamom; or coffee with ginger

In a word
Masha'Allah (God has willed it)

Celebrate this
Yemen's pre-Islamic documented history is distinct from the rest of the Gulf due to the Sabaean Kingdom's control of the lucrative spice trade routes

Random fact
Despite having been unified for centuries, modern Yemen was actually divided into two countries – South Yemen and North Yemen – until 1990

1. A fisher casts a net off the north coast of Socotra island
2. In the Hadramaut region, settlements follow the flow of water beneath Wadi Da'wan plateau
3. Yemeni men wear a traditional dagger, the *jambiya*, the hilt of which denotes the wearer's status
4. Dragon's blood trees are native to the island of Socotra, where they have a lifespan of 1000 years but are threatened by war

1. Batoka Gorge on the Zambezi River is a top spot for rafting adventures

2. Sorting Zambian coffee beans, an important export crop that is noted for its mild flavour

3. South Luangwa National Park is an important habitat for the African wild dog, a charismatic and rare predator that is threatened elsewhere in Africa

Z CAPITAL LUSAKA // POPULATION 17.4 MILLION // AREA 752,618 SQ KM // OFFICIAL LANGUAGE ENGLISH

Zambia

Waters course through Zambia's three great rivers – the Zambezi, the Kafue and the Luangwa – breathing priceless life into vast tracts of wilderness and providing a unified rhythm for a nation of many cultural backgrounds and fascinating traditions. Beneath its soils run riches too, with veins of copper and cobalt fuelling the economy and garnering huge amounts of Chinese investment. Yet mining is confined, and almost a third of the entire country's diverse landscape is reserved for wildlife. As a result, Zambia's national parks offer a variety of phenomenal safari experiences, with some of the world's best guides – walk through South Luangwa, paddle a canoe in the Lower Zambezi or 4WD deep into Kafue. In Zambia visitors don't just see wildlife, they get to know it.

Best time to visit
May to early October

Top things to see
- Victoria Falls, where a million litres of water plunge over the mile-wide chasm per second
- Africa's other great wildebeest migration in the little-known Liuwa Plain National Park
- Leopards and other wildlife in Kafue National Park's 'Emerald Season' (December to April)
- The Lozi king leading his people's retreat from the annual floods in the Ku'omboka ceremony

Top things to do
- Witness African wildlife at eye level while on a walking safari in South Luangwa National Park
- Snorkel in Lake Tanganyika's fish-filled waters
- Raft the Zambezi's Batoka Gorge, home to some of the world's most famous white-water rapids
- Peek over the edge of Victoria Falls from Zambezi's precariously placed Devil's Pool

Getting under the skin
Read: *Kakuli*, a story of wild animals, by Norman Carr; Namwali Serpell's *The Old Drift*, an award-winning Zambian puzzle that spans centuries
Listen: to *kalindula* music; anything from Mampi, Zambia's 'Dancing Queen'
Watch: *I Am Not a Witch*, a BAFTA-winning film about a young Zambian girl
Eat: *nshima* (porridge made from ground maize) with *chibwabwa* (pumpkin leaves) and *nkuku mu chikasu* (village chicken)
Drink: Tiemann Beer's Wild Dog wheat beer

In a word
Muzuhile? (How are you? in Bemba)

Celebrate this
Christabel Mwango: the program coordinator for the non-profit NGO Alliance for Accountability Advocates Zambia, she works to create a well-informed generation of young leaders.

Random fact
The annual fruit bat arrival in Kasanka National Park is the world's largest mammal migration.

Z **CAPITAL** HARARE // **POPULATION** 14.5 MILLION // **AREA** 390,757 SQ KM // **OFFICIAL LANGUAGES** SHONA, NDEBELE, ENGLISH, XHOSA, VENDA, TSWANA, TONGO, SOTHO, SHANGANI, NDAU, NAMBY, KOISAN, KALANGA, CHIBARWE, CHEWA & SIGN LANGUAGE

Zimbabwe

A land of hope and worry, Zimbabwe has yet to find its feet in the post-Mugabe era. Little wonder, given the economic ruin left at the end of his 37-year rule. But this is a nation with a proud past, one Zimbabweans don't have to look far to see – rising in the south is Great Zimbabwe, intricate vestiges of a medieval city whose kingdom dominated Southern Africa and traded gold and goods with places as far away as China and Arabia. Great wealth remains in modern Zim – it's found foremost in its people, but also in its mineral resources and national parks. The latter offer some of Africa's most rewarding wildlife encounters and protect diverse landscapes, which vary from the wonders of Victoria Falls and the Zambezi Valley to the craggy heights of the Eastern Highlands.

Best time to visit
May to October (the dry season)

Top things to see
- The medieval city of Great Zimbabwe, once the religious and political capital of a powerful African kingdom
- Victoria Falls thundering below from the seat of a microlight aircraft
- The sacred Matobo Hills, a dramatic showroom of 3000 ancient rock-art sites
- Endangered African wild dogs hunting en masse in Hwange National Park
- Fossilised skeletons of dinosaurs while hiking on the Sentinel estate

Top things to do
- Share the Zambezi with hippos and crocs while canoeing in Mana Pools National Park
- Breath the cool mountain air while trekking in the Eastern Highlands
- Fly like a superhero across Batoka Gorge on the 'Zambezi Swing'
- Track black rhinos along the edge of Lake Kariba in Matusadona National Park
- Experience the 'wilderness of elephants', otherwise known as Gonarezhou National Park

Getting under the skin
Read: *This Mournable Body* by Tsitsi Dangarembga, which follows the hope and struggles of a young girl in Harare; *We Need New Names* by NoViolet Bulawayo, a story of life under the Mugabe regime
Listen: to Chiwoniso's last album, *Rebel Woman*, a tribute to Africa's women
Watch: *Mugabe and the White African*, a documentary following a farmer's fight to save his land; *Robert Mugabe...What Happened?*, a look at the successes and failures of Zimbabwe's ex-leader
Eat: *sadza ne nyama* (maize-meal with meat gravy)
Drink: Bohlinger, a Zimbabwean-brewed lager

In a word
Mhoro (Hello, in Shona)

Celebrate this
Tsitsi Dangarembga, whose debut novel, *Nervous Conditions*, was picked by the BBC as one of the 100 Novels That Shaped Our World: her latest work, *The Mournable Body*, was shortlisted for the Booker Prize 2020.

Random fact
Lake Kariba, created by the building of Kariba Dam in 1959, holds more water than any other artificial lake in the world (more than 180 cubic kilometres).

1. Victoria Falls on the Zambezi River is bordered by Zambia and Zimbabwe

2. A village near the ruins of Great Zimbabwe, a city built between the 11th and 14th centuries

3. Around 500 lions roam Hwange National Park, some of which, unusually, prey on the park's many elephants

Dependencies, Overseas Territories, Departments & Administrative Divisions/Regions

Australia
- Ashmore & Cartier Islands
- Christmas Island
- Cocos (Keeling) Islands
- Coral Sea Islands
- Heard Island & McDonald Islands
- Norfolk Island

China
- Hong Kong
- Macau

Denmark
- Faroe Islands
- Greenland

France
- Bassas da India
- Clipperton Island
- Europa Island
- French Guiana
- French Polynesia
- French Southern & Antarctic Lands
- Glorioso Islands
- Guadeloupe
- Juan de Nova Island
- Martinique
- Mayotte
- New Caledonia
- Réunion
- Saint Pierre & Miquelon
- Tromelin Island
- Wallis & Futuna

Netherlands
- Aruba, Bonaire and Curaçao
- Saba, Sint Eustatius, and Sint Maarten

New Zealand
- Cook Islands
- Niue
- Tokelau

Norway
- Bouvet Island
- Jan Mayen
- Svalbard

UK
- Anguilla
- Bermuda
- British Indian Ocean Territory
- British Virgin Islands
- Cayman Islands
- Falkland Islands
- Gibraltar
- Guernsey
- Jersey
- Isle of Man
- Montserrat
- Pitcairn Islands
- St Helena
- South Georgia & the South Sandwich Islands
- Turks & Caicos Islands

USA
- American Samoa
- Baker Island
- Guam
- Howland Island
- Jarvis Island
- Johnston Atoll
- Kingman Reef
- Midway Islands
- Navassa Island
- Northern Mariana Islands
- Palmyra Atoll
- Puerto Rico
- Virgin Islands
- Wake Island

Other Places of Interest

The following destinations don't fit neatly elsewhere in this book. They are officially dependencies of other nations but we believe that these places are of special interest, whether that be due to wildlife or history or geography, and they are generally considered to have a strong independent identity and to be quite different from their parent countries. Tony Wheeler, Lonely Planet's founder and perennial explorer, compiled this section.

Aruba, Bonaire & Curaçao

CAPITAL ORANJESTAD (A), KRALENDIJK (B), WILLEMSTAD (C)
POPULATION 110,663 (A), 15,800 (B), 227,000 (C)
AREA 180 SQ KM (A), 294 SQ KM (B), 444 SQ KM (C)
OFFICIAL LANGUAGES DUTCH, SPANISH, ENGLISH, PAPIAMENTU

It's possible that the 'ABC Islands' are the most concentrated area of multiculturalism in the world. Papiamento, spoken throughout the islands, is testament to this fact – the language is derived from every culture that has impacted on the region, including traces of Spanish, Portuguese, Dutch, French and local Indian languages. The islands are diverse: upmarket Aruba is the most touristed island in the southern Caribbean, Bonaire has an amazing reef-lined coast, and go-go Curaçao is a wild mix of urban madness, remote vistas and a lust for life.

Montserrat (UK)

CAPITAL PLYMOUTH
POPULATION 5241
AREA 100 SQ KM
OFFICIAL LANGUAGE ENGLISH

In 1997 a volcanic eruption devastated Montserrat. Despite plenty of warning that after 400 sleepy years the Soufrière Hills Volcano was about to wake up, there were still 19 deaths and the capital, Plymouth, was buried. Reconstruction is still going on over two decades later, but despite this there's a certain magnificent desolation about the place. Montserrat's few visitors come for volcano-related day trips. Those who stay longer will relish the solitude and enjoy the chance to become part of the island's rebirth.

Niue (New Zealand)

CAPITAL ALOFI
POPULATION 1190
AREA 259 SQ KM
OFFICIAL LANGUAGES NIUEAN & ENGLISH

Midway between Tonga and the Cook Islands, which makes it a long way from anywhere, Niue is an example of a makatea island, an upthrust coral reef. It rises often vertically out of the ocean so there's very little beach, but in compensation there are amazing chasms, ravines, gullies and caves all around the coast. Some of them extend underwater, giving the island superb scuba-diving sites. Like a number of other Pacific nations, the world's smallest self-governing state has been suffering a population decline: today there are more Niueans in New Zealand than on 'the Rock of Polynesia'.

Réunion (France)

CAPITAL ST DENIS
POPULATION 844,994
AREA 2517 SQ KM
OFFICIAL LANGUAGES FRENCH

Réunion is so sheer and lush, it looks like it has risen dripping wet from the sea – which it effectively has, being the tip of a massive submerged prehistoric volcano. The island is run as an overseas department of France, and French culture dominates every facet of life, from the coffee and croissant in the morning to the bottle of Evian and the carafe of red wine at the dinner table. However, the French atmosphere of the island has a firmly tropical twist, with subtle traces of Indian, African and Chinese cultures.

South Georgia (UK)

CAPITAL GRYTVIKEN
POPULATION 10-20
AREA 3755 SQ KM
OFFICIAL LANGUAGE ENGLISH

Looking like an Alpine mountain range soaring out of the ocean, South Georgia's topography is matched only by its spectacular wildlife. The island's human population, scientists of the British Antarctic Survey, may drop to 10 during the winter. But there are two to three million seals, a similar number of penguins and 50 million birds, including many of the world's albatrosses. Add industrial archaeology in the shape of abandoned whaling stations, plus South Georgia's role in Sir Ernest Shackleton's epic escape from the ice, and it's no wonder this is a popular destinations for tourists.

Svalbard (Norway)

CAPITAL LONGYEARBYEN
POPULATION 1872
AREA 61,229 SQ KM
OFFICIAL LANGUAGE NORWEGIAN

Far to the north of Norway, the archipelago of Svalbard has become a popular destination for Arctic travellers, keen to cruise the ice floes in search of whales, seals, walruses and polar bears. Apart from wildlife there are also some terrific hiking possibilities on the main island where you might encounter reindeer and Arctic foxes. The main town, the engagingly named Longyearbyen, has a long history of coal mining.

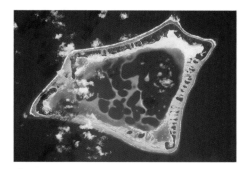

Tokelau (New Zealand)

POPULATION 1337
AREA 12 SQ KM
OFFICIAL LANGUAGE TOKELAUN,
ENGLISH, SAMOAN

Tokelau consists of three tiny atolls, each of them laid out on classic atoll design principles: a necklace of palm-fringed islands around a central lagoon. Off to the north of Samoa, the islands are not only a long way from anywhere, but also from each other; it's 150km from Atafu past Nukunonu to Fakaofo. They're also very crowded: there may be less than 1400 people but none of the islets is more than 200m wide and you've got to climb a coconut tree to get more than 5m above sea level. Getting there is difficult even for sailors, as none of the lagoons has a pass deep enough for a yacht.

Tristan da Cunha (UK)

CAPITAL EDINBURGH
POPULATION 267
AREA 98 SQ KM
OFFICIAL LANGUAGES ENGLISH

Officially a dependency of St Helena, over 2300km to the north, Tristan da Cunha is frequently cited as the most remote populated place in the world. The island is a simple, towering volcano cone, and an eruption in 1961 forced the complete evacuation of the island. The displaced islanders put up with life in England for two years but most of them returned as soon as the island was declared safe in 1963 and went straight back to catching the crawfish that are the island's main export. Nightingale, Inaccessible and two smaller islands lie slightly southeast of the main island.

Wallis & Futuna (France)

CAPITAL MATA'UTU
POPULATION 15,613
AREA 274 SQ KM
OFFICIAL LANGUAGES FRENCH

This French Pacific colony is made up of two islands, separated by 230km of open ocean and quite dissimilar. Wallis is low lying with a lagoon fringed by classic sandy motus (islets), while Futuna is much more mountainous and paired with smaller Alofi. The populations are equally dissimilar: Futuna has connections to Samoa while the Wallis links were with Tonga. Wallis has one of the Pacific's best archaeological sites at Talietumu and an unusual collection of crater lakes, while both islands are dotted with colourful and often eccentrically designed churches.

THE TRAVEL BOOK

A JOURNEY THROUGH
EVERY COUNTRY IN THE WORLD

4th Edition
General Manager, Publishing Piers Pickard
Associate Publisher Robin Barton
Editor Bridget Blair
Update authors Bradley Mayhew, Brana Vladisavljevic, Brett Atkinson, Cliff Wilkinson, James Smart, Joe Bindloss, Jess Lee, Matt Phillips, Sarah Stocking, Tamara Sheward
Design and Pre-Press Production Lauren Egan
Print Production Nigel Longuet
Cover illustration Whooli Chen

Published in 2021 by
Lonely Planet Global Limited
Digital Depot, Roe Lane (off Thomas St),
Digital Hub, Dublin 8, D08 TCV4, Ireland

Stay in touch lonelyplanet.com/contact

ISBN 978 1 8386 9459 3
10 9 8 7 6 5 4 3 2 1
Text & maps © Lonely Planet Pty Ltd 2021
Photos © as indicated 2021
Printed in Malaysia